More Praise for *Leaving Resurrection*

"What did it mean?" Eva Saulitis writes at the very start of *Leaving Resurrection*. As she well knows, hers is the question of a lifetime and in this wonderful essay collection, she sets about seeking her answers. Her explorations take her through the waters of Prince William Sound hunting killer whales, but that hunt driven by the needs of scientific research is only the starting point for deeper and more profound hunts: for the nature of the human heart and spirit. In her search, Eva Saulitis neglects nothing, from the bull kelp washed onto the beach, to the abandoned military gun placements above Resurrection Bay, to the varied behavior of the whales she follows. Mostly, though, she probes herself for the hardest questions, the most elusive answers.

Eva Saulitis is a fearless hunter. Her best tool is the rich music of her language which serves her scientific and poetic self. Like the whales, she sings mysterious songs to her readers, leading us on to new places, leading us off the map of the familiar. That is her goal, "To create a new model: that challenges all of my knowledge, all that I've experienced, all the questions I've learned to ask." These essays are that model, a fusion of head and heart, a rich wonderment, an invitation to a deeper understanding of the world and of ourselves.
—Frank Soos, author of *Unified Field Theory*

Eva Saulitis is both a scientist and literateur, a formidable combination that makes the strength at the core of the essays collected in *Leaving Resurrection*. She has the scientist's precision when recounting natural phenomena, notably Alaska's Prince William Sound and the creatures that inhabit it, yet what gives legs to the precision is wondrous writing and the spiritual and conjectural reach of the memoirist. There is a lot in this book, including the memory of childhood, love, abiding friendship, and thoughtful, intimate, sometimes chilling accounts of killer whales, and even arresting tales of hazard at sea that are sure to make the reader's muscles twitch. This book gets better and better the deeper one goes into it, and so, too, its amplitude and complete logic intensifies, resonating after the last page is turned.
—John Keeble, author of *Out of the Channel*

Eva Saulitis is that rare blend of poet-philosopher and scientist, akin to John Muir. Like Muir, she embraces both rigorous inquiry and spirited passion in her quest to understand, broadly, the natural world that surrounds and connects us. *Leaving Resurrection* is a deeply searched and beautifully written account of a storied life given to finding a way from data and information to something like wisdom.
—Nancy Lord, author of *Beluga Days: Tracking the Endangered White Whale*

Leaving Resurrection

Eva Saulitis

with illustrations by Karl Becker

Boreal Books | Fairbanks, Alaska

The poem "Earth Verse," by Gary Snyder, from *Mountains and Rivers Without End*, is reprinted by permission of Counterpoint Press, a member of the Perseus Books Group.

Seven haiku from *The Essential Haiku: Versions of Basho, Buson & Issa*, edited and with an introduction by Robert Hass. Introduction and selection copyright © 1994 by Robert Hass. Unless otherwise noted, all translations copyright © 1994 by Robert Hass. Reprinted by permission of HarperCollins Publishers (Ecco Press).

Cover art: Detail from *I divide, I divide, I divide* by Marissa Favretto , 2002. Oil on canvas.
Book design by Mark E. Cull

ISBN-13: 978-1-59709-091-9
Library of Congress Catalog Card Number: 2007940745

Boreal Books is an imprint of Red Hen Press
First Edition

for Molly Lou

for Craig

for Elli, Lars, and Eve

&

for the place
and the animals
that brought us together

Contents

—and you, my memory, do not prevent this afterlife,
something told to somebody, then retold and changed, and so forth—
conversation becomes a shifting scene, and above a current of questions
how shall I be, of silt, of tide—?

—Molly Lou Freeman

Kinds of water drown us; kinds of water do not.

—Anne Carson

Preface

In the summer of 1986, I flew by way of a De Havilland Beaver and a wild-eyed Vietnam vet pilot into Prince William Sound, Alaska, on my way to my first post-college biology job at a fish hatchery. As the pilot dive-bombed friends on fishing boats, their seine buoys bobbing on the whitecaps, I already longed for that place. I lived in Prince William Sound that year and have returned every summer since to study killer whales, first for a master's degree, and now as part of an ongoing research project. For most of those years, I migrated every winter to Alaska's interior, to Fairbanks, a place of extremes, where four hours of blue light illuminate winter days, where fifty below temperatures slice the nights into shards. Even the moose walk carefully in that cold.

According to theologian Peter Gomes, Celtic mythology describes "thin places" on the earth, where the material and spirit worlds exist in close proximity. The Celts had their sacred groves. The Chugachmiut people in Prince William Sound had their mummy caves. As a child, I found protection under a quaking aspen sapling growing in the middle of a spruce forest my father planted.

These essays are set in the thin places. As a young biologist learning to navigate inner and outer terrains, I came up against the unexpected encounter, both scientific and personal. Even while I jotted notes and observed killer whale behavior in objectivity, the story-making self, the part watching me watch the whales, jotted unscientific notes, asked unscientific questions. Science requires a detachment from parts of the self, requires freedom from subjectivity, from bias. One tries to see things clearly, as they are, apart from one's prejudices and emotions. But living a whole life in this way is unsatisfying, so in my field camp, I stayed up until 3:00 a.m. in the perpetual light of summer, writing, trying to put my whole self and perception back together. Perhaps, for some, that kind of writing calls into question my validity as a scientist. But I believe that this weaving and constructing can cohabit with the self-effacement necessary to be a scientist, that the problems of this strange marriage are real for others, and grow ever more pressing in our relationship to technology and the earth. These struggles belong to us all:

our responsibility toward what we study, the question of detachment, striving for and against it, which is the striving for truth.

The essays that follow are transparencies, thin places, divided, stitched back together, an ugly cloth left out in the rain, an old wall tent's hide, sewn and patched with oil-soaked pieces—stories—smelling of fumes and salt and air-dried canvas, both musty and fresh. One eye looking out at the world, one tarped body, wrapped in narratives, one brain asking *what does it mean*? It is one way of seeing, this pile of bones on the beach, this small, still-warm corpse in the hand.

Leaving Resurrection

The Burden of the Beach

What is it to breathe not of this air?

—Molly Lou Freeman

The killer whale is long gone now, washed off the beach where we found it. By the roiling of stones, the ocean has erased evidence of its weight, the oil and blood, even the bones. It breathes a different air, the air of memory. But our question remains: *What did it mean*?

A carcass is an inevitable thing for a biologist. We can ask it why, and it will lie there, a puckered eye looking up, deaf even to the simplest question. How did it get here? Why did it die? Or we may never see it, grounded on a remote pocket beach, visited only by gulls, eagles, ravens, bears.

Our gum boots slip on algae-slick stones as we approach the whale. Hesitating every few feet, we call out, "Halloooo, bears . . . coming throooough bears . . ." The southwest sea breeze carries our shouts across the tiny isthmus. Setting our buckets down, we walk in opposite directions, circling the carcass. Molly Lou stops near the dorsal fin.

"God, it reeks!" she grimaces. We laugh, and I think, *Can we do this*?

We came with specific questions. What killed this whale? What population does it belong to? What did it eat? We came to find this body, measure it, photograph it, cut into it, collect its blubber and skin, remove its stomach and take what's inside. The dead whale doesn't instruct us. It roasts in the sun, cracked and dripping grease, skin no longer black and white but rose-streaked bronze.

I scan the tall grass at the top of the beach, the forest's shifting screen. Brown bears might watch us from those trees. Once, as we tracked a group of killer whales in our boat, *Whale 1*, along a beach not far from here, a brown bear lumbered near the water's edge, parallel to us. The whales had been down for a few minutes. When they surfaced again, the bear startled and stood up, staring.

"Hallooooo bears," I cry.

Parts of this island are thick with brown bears, which inhabit only three islands in Prince William Sound. Opening like a bear's wide gape to the Gulf of Alaska, ten-mile-wide Hinchinbrook Entrance separates the two largest islands, Montague and Hinchinbrook. A dangerous strait, it's streaked by currents and riptides. Winds rise unexpectedly. Through its mouth, storms enter the Sound, whirling off the Gulf.

A brown bear once swam across the Entrance, a crossing we dare not make in our twenty-foot boat. She'd been raiding garbage cans in town, so biologists transplanted her forty miles away to Montague Island. A few weeks later, she was back.

Chugachmiut people of Prince William Sound tell of bears swimming in the opposite direction. Once, at a now-abandoned village on Hinchinbrook Island, a woman, betrayed by her husband, transformed herself into a brown bear and killed him. Four fur seals disguised as men bore her by baidarka to Hinchinbrook Entrance. From there she swam into the Gulf of Alaska, creating Middleton Island out of seaweed so she could rest. She finally arrived at Montague, her fury intact. That's why, it's said, brown bears on Montague Island are so fierce.

I imagine the betrayed woman's changeling descendants, watching us through the rustling alders. Surely the wind carries the whale's stink through the forest, but maybe this carcass is too rotten, even for bears. Or maybe, like humans, bears shun the flesh of killer whales. "Let's get this over with as fast as we can," I say, opening a bucket lid. "Let's make a plan." I hand Molly Lou a knife.

"Eva, why do you think we have to do this?" she asks.

As usual, Molly Lou stops me with a question not on our scientific agenda. I want to ponder it, but I'm in science mode, and I'm nervous. I've never done this. "I don't know, Molly," I say, but the question loads itself like a roll of film in my mind.

On the boat, we'd changed into disposable clothes we bought at a second-hand shop in Valdez and filled five-gallon buckets with new fillet knives, notebooks, and plastic bags. We didn't have time to track down flensing lances, gauntlet gloves, chain saws—the usual tools used by scientists and scavengers for cutting into whales. And we didn't have a gun. Guys in Valdez told us we were crazy to be tearing into a whale carcass on Montague Island without a shotgun. All we have is our voices, so we talk incessantly. We tease one another about our tight designer jeans, my pink sweatshirt, the arms too short, Molly Lou's tourist t-shirt. As we pull on heavy raingear, Molly Lou quotes Adrienne Rich: "I pull on the body armor of black rubber . . ." The poem is called "Diving into the Wreck."

I examine the wreck of the whale. Killer whales don't seem this big from the water, even when they swim right under our boat. This one's twenty-two-feet long, two feet longer than *Whale 1*, according to our measuring tape. When Molly Lou walks to its dorsal fin, I see only her head. I pull out a fillet knife from the bucket.

"Maybe one of us should watch for bears, while the other cuts," she suggests. I agree, looking back down the beach to our inflatable raft, and then to the *Whale 1*, at anchor fifty yards offshore. I imagine scrambling to the raft, paddling like mad for the boat, a chuffing bear in pursuit. At least with the receding tide, the boat's getting closer. I turn, and knife in hand, press its tip into the crisp throat skin of the whale.

It's not just bears. Legends say this island's haunted, and it seems designed to keep people away. During the 1964 earthquake, parts of the Montague Island shoreline rose thirty feet. The charts, not yet revised, warn of unmarked reefs and ledges. At low tide, strange swirls suggest these dangers.

With its permanent snowfields and pocket glaciers, the island's forty-mile jagged backbone bears along its length the Gulf's force. Storms pile up on the ocean side. Weather meets the mountains and boils down the leeward slopes like foam. In some valleys, more than two hundred inches of rain fall in a year.

On nautical charts, the tiny peninsula where we stand looks like a beckoning finger. I've only been ashore on Montague a few times, though from the water, I've stared at it like an unrequited lover, imagining myself wandering in groves of old growth, hiking up broad green valleys, crossing big rivers, skiing the smooth snow bowls, disappearing over the ridgeline to the mysterious other side.

Earlier in the summer, friends and I landed a skiff on the island. While they hiked, I crouched, writing, in a rye-grass meadow strewn with beach logs tossed up during winter storms. I didn't sit facing the water as I usually do, but facing the island, so I could see what was coming. I wondered where the closest bear was, imagined its eyes in the spaces between leaves, its tongue stripping berries from branches. I felt strangely relaxed and thought of old Tlingit stories of women who coupled with bears. I wanted a bear to come to me.

I sat up suddenly, my skin prickling. *What's the matter with you?* I hissed out loud. Glancing around, I gathered my diary and pack and walked quickly back to the beach, wondering what I'd really wanted. To be taken? To be claimed?

It's hot. Under my raingear, my body sweats as I rip the knife through the whale's rubbery skin. Because it stinks so much, I stand an arm's length from the carcass, face turned away. As the knife enters the blubber, juice oozes out. I feel its warmth through my glove. In the whale, tiny organisms break down tissues, cell by cell. I saw away a square of skin and melting fat and drop it into the bucket Molly Lou holds toward me. A lab will whirl this flesh in test tubes, and by a process called gas chromatography, render it into data on contaminant levels in the whale.

Nothing smells as bad, the thick, fermented sweetness, the pungent aftertang. In my freezer at home, parts wrapped in ten layers of plastic still taint Ziplocs of berries and packages of fish with whale stink. Some biologists conducting necropsies on whales smoke cigars.

Molly Lou paces the pale cobbles, shouting and singing. "How are you doing?" she asks.

"It's so hot. And this smell . . . I'm trying not to judge it. If I describe it to myself, it's not so bad." I turn my face, gulping breaths of wind.

She offers to start excavating for the stomach, one of our primary tasks, besides getting skin and blubber samples. In the Pacific Northwest and Alaska, two ecotypes of killer whale coexist, fish-eaters and mammal-eaters, but many scientists are still incredulous. Historically, killer whales were considered, at best, opportunists, and at worst, ravenous killing machines. An infamous paper from the '50s lumps known killer whale prey into a list that includes everything from herring to squid to bearded seals to blue whales. But each stomach, examined separately, tells a different story, that the killer whale is not one, but many beasts, that each population has a unique diet that changes seasonally. Even individual killer whales may specialize on certain prey. The story, as one digs deeper, grows more, not less, complex. And Molly Lou and I know, right now, we're part of that story.

I hand Molly Lou the gloves. We stand side by side. "Let's just cut it like we're gutting a fish," I suggest, trying to visualize where the stomach will be, above the chewed-open flesh near the belly button and below the chest. The giant, paddle like pectoral flippers fold together, blocking access to the upper belly.

"Let's just lift the top flipper out of the way," says Molly Lou, striding up and grasping the peeling slab. I join her, and we both heave. It doesn't budge; the flipper could be made of cement. Bending down and crawling as far as she can under the flipper, Molly Lou begins to cut.

Several minutes have passed since we've shouted, so I call out as I walk around to the whale's head. The flesh puckers where gulls picked out the skyward-facing eye. The mouth gapes, revealing rows of evenly spaced, conical teeth. I bend down and stroke one, find it cool and smooth, harder than bone, more like marble.

I should cut one out for aging, I think and fetch our hacksaw from the bucket. After several minutes of vigorous rasping, I let my arm drop. The tooth is barely scratched. The hacksaw, like the knife, is nothing to the whale, a joke. I leave it and walk back to watch Molly Lou.

She emerges from under the flipper, jerking the knife along, jaggedly tearing through the blubber. On her knees, wallowing in her stiff raingear, she's bent at an awkward angle, her face hovering inches from the whale. Blood and grease blobs smear her rubber-suited arms. Split-open rinds of blubber fall away from the underlying meat.

I'm amazed at her willingness, her endurance. She's twenty-five, six years younger than I am, five-two with a delicate frame, and normally, she's fastidiously clean. But she grew up on the Alaskan coast, on a homestead without running water. From early childhood, she crewed on her father's gillnet boat, *Sweet Sage*, fishing on turbulent, tide-swept salmon grounds. I grew up in a ranch house in rural western New York. My family never went camping or hiking. For work, I stocked grocery store shelves and picked grapes. Nonetheless, we've become sisters.

We were strangers when we met five years ago, but intimacy quickly grew through the four-month-long field seasons, sharing a wall tent on an island thirty miles from here, seventy miles from the nearest town. She's been my field assistant, helping me to collect data for my master's thesis on killer whale behavior and vocalizations. We search the passages for killer whales, often days passing without them. Sometimes we talk all day, about our families, regrets, lovers, struggles, books, poems. We tell jokes, make up words and sing. Sometimes we question each other too much, digging into hard places with our excavating tools. Sometimes hours go by in silence. Then, I'll begin speaking in the middle of a thought, and it won't seem to matter.

We hike every day, stretching our legs after hours on the boat, but also as a kind of ritual. We explore beaches, the smallest islands, deer trails up ridges, bogs where we walk barefoot, ankle-deep in moss and water. We challenge each other to swim in every kind of weather. Along the shore, we search for Japanese fishing floats, colored glass balls that survive wave-driven landings. On the same foggy day last summer, we found our first glass balls on two separate beaches. In the evening, we lounged in the wall tent, throwing coal chunks into the stove, holding our glass balls to the light, peering through them. Molly Lou's was small, fitting perfectly in her palm. The glass was a dark, clear green. Mine was larger, the pale blue glass opaque, scratched.

"They're like us," I said. "Yours is so clear-sighted and compact; mine is cloudy and uncertain, a little battered." We laughed.

"I think yours is beautiful," she said.

Next summer, Molly Lou won't be here. She's a poet, not a scientist. She needs to work in her own field—teaching and writing. She brought poetry to our work. She taught me the question of poetry, which is also the question of science, *What does it mean?* I don't know, anymore, if I can do science without that balance. How will I do it, I wonder, live in this intimate, honest way, with another scientist as my field assistant, with a stranger?

Molly Lou lurches up, her face flushed. "I have to get away from this and breathe some fresh air," she says, handing me the gloves and knife before stumbling down the beach to move our buckets closer to the water's edge. The tide has receded several feet. "Heeeeyooo . . ." she yells over her shoulder.

As I slice at the brown, fibrous meat, juice sizzles out. After death, a whale's body holds heat for days, cooking inside while the sun warms it from without. Once, muscles contracting, tail fluke pumping, pushing water up and down, this being glided through the sea. It swam a hundred or more miles a day, hunting salmon schools or twirling porpoises. When a body's cast out of its element, it sags, collapses, breaks down.

I see myself suddenly with my knife poised over the carcass. I, too, am out of my element. I'm embarrassed, as if something unseen scrutinizes me. Death, for most animals, is private, secret. What is this efficient person, scanning with binoculars, wielding cameras and knives, scratching notes on wet paper with a mechanical pencil? Sometimes, watching the distant mountains of Montague, the streaked rock lenses eddied by ice and snow, I think, *I want my mind to be like that.* Mountains reflect me upside-down. On the boat, I peer down into the water, straining to see what's beneath: the eye of a whale, or its shoulder, a flash of black and white in the green.

Two hours have passed. "We should go faster," I tell Molly Lou. "Maybe we can both cut and yell at the same time."

We carve away rinds of fat and muscle to expose a rectangle of smooth, gray tissue—the abdominal cavity's taut membrane. When I push the knife tip through, a whoosh of putrid air rushes out, and the membrane collapses. Jumping back, we hold our breath, turning our heads away. "Jesus," says Molly Lou, gasping. When I slice away a patch of lining, glistening, veined intestines coil out. My knife slits a loop, and it oozes orange slime, the color of sulfur fungus.

The stomach is deep inside the whale. To make space, I stand back and pull intestines onto the beach. We take turns grasping and probing. Finally, I feel a widening flask. Bent under the ribs, lost in a mass of intestines, my arms extend as far as they can into the whale's body.

"How the hell are we going to get this out?" I ask.

"We need a bigger opening," Molly Lou says, knifing off more slabs of blubber and muscle, widening our rectangle. I wield the hacksaw, trying to carve away ribs. As with the teeth, the saw just scratches the two-inch-thick bone.

We decide to climb on top of the whale and heft the stomach up from under the lowest rib. Using the pectoral fin as a step, Molly Lou crawls across the whale's side while I cut away transparent membranes holding the stomach pouches to the abdominal wall. When a mound of stomach loosens, I lean back and haul the slippery mass out from under the rib and heave it to Molly Lou. Bending her knees, she strains upward until the sacks drape over the whale's side.

When I try to climb up to help her, my foot slips on grease, and I fall into the whale. Shin-deep in blood and body fluid, I look up beyond Molly Lou's shocked face to a cadre of eagles and ravens regarding us from the trees. I try to imagine what they see, two orange-suited creatures crawling all over the carcass, cutting, probing, shouting. Do they wonder, what is the purpose, what place do these animals occupy? Once I stared over the bow of the boat at my reflection in the water. *This is what the whale sees, looking up*, I thought, an odd, flickering, white-faced shape flapping its orange wings, jabbering in the blinding air.

"Are you okay?" Molly Lou asks. Giving me her hand, she hauls me out of the whale, and I climb back up, crawling across to her.

It takes both of us pulling, but soon the stomach is heaped on the whale's side. Kneeling down, we draw our knives back and forth across pouches. The blades are dull from cutting so much fiber and muscle. When I finally slit through, the stomach's inner surface is white and crenulated, like the underside of a mushroom. This pouch is empty.

"I think I've found something," Molly Lou says quietly. Her palms press on a bulge, hard and knobby, like a bird's crop. When she cuts it, a brown soup of whiskers, claws, and shriveled skin strips spills out. We pull out handfuls of quill-like sea lion whiskers and the white, pearly ones of seals.

Something odd pokes from the mass. I reach into the jumble and pull out three square tabs. "What the hell . . ." I begin. They look disconcertingly like buckles from raingear. Then I recognize them as numbered flipper tags that biologists somewhere snapped to living seals or sea lions, to trace their movements and fate.

Excited now, we open more pouches, scraping the contents out with our fingers, filling Ziploc bags, not wanting to miss tiny fish scales that may cling to the stomach walls. We eventually find fourteen flipper tags. When we finish, we slide off the whale, our boots crunching down on the cobbles.

Dripping slime, we run to the water's edge and splash our raingear to clean it. "We did it!" I shout. When I turn back, the thing on the beach deflates my exhilaration. Streaks of intestine spill from the hole, stick to rocks. Brownish fluid fills the cavity, where organs sag.

I remember a radio call we received last summer, reporting a headless killer whale floating in Montague Strait. Disgusted, I imagined a chain saw–wielding someone, after the teeth or the skull. But how was this person different from me? Both of us, wanting something, and taking it, from the whale.

Some traditional Alaska Native people perform rituals when they kill something. Some slit the eyes of dead seals, so the animals can't see their own bodies butchered. There are some animals, like killer whales, they do not touch. I don't know what ritual will make this right. This is it, the paradox we—scientist and poet—live out.

Back on the boat, we prepare to leave. Molly Lou dunks a five-gallon bucket over the side. Still in our raingear, we stand on the bow and squirt dish soap, then douse each other with water, scrubbing away offal with long-handled deck brushes. After several washes, the smell remains. We imagine it will never leave. We peel off our damp clothing and pile it in a plastic bag for later burning. After soaping our bodies, we take quick dives into the water, but the whale's scent has permeated our skin.

Finally, in dry clothes, I pull the anchor and Molly Lou backs the boat away from shore. The tide's begun to rise again, creeping up the beach.

I remember a place we found last summer, a green slope thick with salmonberries, which we ate as we climbed. On the ridgetop, we jumped into a cold pond, then fell asleep in a meadow. Afterwards, we stood on the beach, looking back toward the slope. Side by side, our palms pressed together, we bowed. We knelt on the beach and touched our foreheads to the wet gravel before we turned and walked away.

Distance erases the wounds in whale's body, but not the memory of it, of what we found there, where two currents met, the scientist and the poet asking why. On a wooden support inside our wall tent, I pinned a quote by Rilke, who said that we should live our questions, rather than striving so hard for answers.

The dredging of the whale occurred in a single afternoon, so its reality remains elusive, like its death, secret still. But we'll keep living our question. Finally, the question, like the mountains on Montague, will become our life. It will be like honesty, necessary and difficult. It will become poetry.

Even as we, hesitatingly, make the first cut in the whale's face
its flesh is so thin here,
we are the fragile ones.

It will one day tell us part of the whale's story, that it was contaminated by DDT and PCBs, that it fed only on mammals, that it belonged to a tribe of killer whales that spends most of its time in the Gulf of Alaska, beyond the protection of Montague Island. That it ate fourteen Steller sea lions tagged by scientists on Marmot Island. But it is only a beginning, an entering into the whale's decomposing, wading through decay until the gentle articulations of bones are revealed.

Even the island holds the whale
by query . . .

We turn our boat south, make for Point Helen.

Behind us, the whale lies on the island's wild fringe, eagles perched in the trees above it. The whale imprints its memory onto the beach. Bowing, we leave the bruised form to waving grass, birds, and bears. I imagine them biding time, waiting to reclaim what's theirs, their eyes in the alders, watching.

Walking on Carlson Lake with Bill

The lake is always sheer as itself,
true as a thought and extending down and above:
transparent ground, icicled and crevassed. Smooth
galleries spin underneath us, round as worlds.
There a crack is a question, a weed a confession
or a roof, a stone is a floor and advice.

—-Molly Lou Freeman

It was mid-March, and spring wouldn't come. The sun glared bright and false in the ten-below air, and the wind blew for days on end. Undaunted, my friend Bill and I planned our trip to Fred Rungee's homestead on Carlson Lake, in the Mentasta Range, five hours south of Fairbanks.

In March in interior Alaska, it's not the spring of birds returning and snowmelt and balminess, but people do begin to hope for mellowing temperatures in the face of the sun's steady ascent, the longer days. When the temperature stays cold like it did that March, when the wind barrels down from the Arctic, the sun is almost harsh in its light, in its refusal.

As a child, I felt apologetic in my prayers to God. I imagined millions of beseeching whispers floating heavenward, little hands tugging at the hem of God's robes. Here in Alaska the sun carries that burden of hope, of longing in winter and gratitude in summer. During spring, we receive the sun like a benediction. During the seven months of winter, in the cold and darkness, we feel abandoned. Our belief is tested. The funny thing about nature is that it gives in strange ways. To have expectations of it here is foolish since, in nature, desire and fulfillment flow like a river and a glacier, side by side, one quick and one long.

I thought I knew what I wanted from my trip to Carlson Lake. I wanted to take Bill there to meet Fred, another old-timer, and I wanted to listen to them tell stories in Fred's log house. But the lake got in our way. Looking back, I envision the ice lying in wait, even as it gleamed indifference, even as it held reflections of mountainsides and sky, only that.

When I imagine Carlson Lake now, I feel its necessity. No one drowned there. But when we stepped onto the ice, and I saw the intersection of the lake's life with ours, I was changed. Now the lake is a lens suspended inside me, one I see through.

"Someday, I want you to meet my friend Terwilliger," Bill said. "He must be in his nineties now, still living in Tok. My, he can tell a story. I remember one, began something like this . . ." Bill cocked his head, squinted, as if calling in Terwilliger's voice from a corner of the room. Crimping the skin between his long, silver brows, he tucked down his chin and lowered his voice: "Well, I woke up the next morning, and the old man was dead." A beat, and then Bill's face collapsed into laughter, into his own face and voice. "Wouldn't that be the beginning of a great story?"

On the way to Carlson Creek in the car, Bill told me stories and played reels and jigs on his harmonica, his features reading like a map of the wilderness, seventy-six winters drawn in. His music, too, was a map. The harmonica riffled like a stream I once crossed, Troublesome Creek, a ribbon of water hobbling over a stony bottom as it swept along. Bill's eyes, silver gray, glitter like that, with trouble, when he plays.

The drive to Carlson Creek took us south to Delta, then onto the Alaska Highway to Tok, down the Tok cut-off toward Slana. Though he didn't mention it, I knew that Bill hadn't been to Tok in almost fifteen years. In early March, seventeen years ago, his eighteen-year-old son Billy had killed himself there.

I've never asked Bill about Billy's death directly. Bill and I reveal ourselves to each other the way the image of the sea floor reveals itself when I watch for the bottom from a boat: first there's a whitening, and then, flecks and shadows coalesce into actual forms: kelp fronds, fanning bodies of fish, eelgrass.

A year ago, Bill came over for breakfast, and instead of his joking, singing, yarn-spinning self, he slumped in the rocking chair by the woodstove. I noticed the heavy sprawl of his broad, bare feet. I sat beside him, and he said, "It's the anniversary of my brother Dave's death. You remind me of him, you know, you would have liked him. Now *he* could tell a story. He had hundreds of stories and songs memorized. It hurts." Bill clenched his fist against his chest. A regular actor in the local Shakespeare troupe, Bill embodies the theater's dual masks, the comic—a *joie de vivre* almost manic in its intensity—

and the tragic, equally intense, equally fragile. I held my breath as we bobbed in that moment's glass boat. A word might have sunk it.

Later, when I hugged Bill, I felt solidity in the span of my arms across his back, delicacy in the birdlike feel of his bones. I felt his holding on and the weight of him planted firmly on the ground. I felt the hoarse break in his indrawn breath and then the long letting go. My own losses hovered there too: the recent suicide of my friend, problems in my marriage, depression. We paddle into each other's pain and back away again. Silence is a lake we float on, casting separate lines. A story is the creature unexpectedly breaking the surface.

One day last winter, Bill and I drove down the road in his tiny, white shell of a car, our breath icing the windows. "Ten years ago, I was driving to town in this car; it was right about here, as a matter of fact," he said. "I was listening to Karl Haas's program on public radio. He played a sad piece by Brahms." Bill lifted one hand off the steering wheel and conducted as he sang out the notes to me: "Bomm, bomm, bomm, bomm . . ." He told me how sobs came then, for the first time, for Billy. He wrote a letter to Karl Haas, thanking him.

As I listened, I stared at his big hands, at their waxy, speckled skin, the way the skin thinned—like dried seaweed stretched between stones—when he gripped the wheel hard.

Still two hours from Carlson Lake, in Tok we stopped at the general store for coffee. Gusts banged signs against the metal siding. As we pushed open the door, ice fog billowed in ahead of us. When I came back from the bathroom, Bill leaned on the counter paying for coffee. "Yeah, she still lives there," a large, sallow-faced woman behind the counter answered. "No, I've never heard of them. I've just been here five years." Oblivious to her indifference, Bill's face was alight recalling old times.

Back on the road, Bill was quiet. The light was fading, and I knew that we wouldn't reach the trailhead until it was dark, with a four-mile ski ahead of us in the cold wind. "I used to hate that last part of the drive into Tok," Bill said. The harmonica thrummed lightly, a ribbon of something like smoke seeping from its square openings and coiling around us in the car. It was loss, and memory, and the tree-stippled hills under snow hurtling past.

A half-moon rode high in the sky by the time we arrived at Carlson Creek. After changing into ski boots, I pushed the car door open against the wind. By the light of headlamps, we packed our sleds. At ten below, the wind instantly sapped warmth from my fingertips. By the time Bill had his thin ski boots on, his feet were cold. "My circulation isn't what it used to be," he said, pulling his mukluks back on.

"Yeah, right," I laughed. At seventy-six, Bill goes barefoot to the wood pile in the snow, bikes four miles a day at fifty below zero to get his mail. While Bill tied his knapsack to his sled, I jammed his ski boots into the top of my backpack and strapped his skis to the sides. I knew that the trail was hard packed from snow machines, as fine for walking as for skiing.

Bill and I crossed the road and walked along the shoulder until we saw the opening in the trees—the trail to Carlson Lake. I'd only been to Fred's cabin on the lake once before, two years ago, and that time I'd arrived with friends at two in the morning, Fred waiting up with hot chicken soup, the darkness inside the rambling log house swallowing the kerosene lamplight, except for the puddle over the kitchen table where we sat.

Bill had heard of Fred, but they'd met only once, two decades ago, when Fred showed up, laden with candy for the local kids, at a Christmas party in Tok. Fred, also a storyteller and musician, is in his seventies now, living out on the lake alone. On his antique upright piano, he bangs out Scott Joplin, Al Jolson, Beethoven, and his father's own love songs published in the '20s. I like to turn the powdery yellow pages of his sheet music. Bill can play just about anything, whether it was meant to be an instrument or not. I'd wanted to get the two of them together for a long time.

In the forest, we couldn't feel the wind. The moon cast a Braille of shadows on the snow. I clipped into my skis and shuffled next to Bill. After a mile, we came alongside the edge of a fifty-foot cliff, a place I recognized even in the dark: that bit of overflow from Carlson Creek. My skis skittered across the curve of rippled ice. When we reached the snow surface again, we stopped and stood quietly, listening. Broken-off trees tilted in the middle of the overflow. They could have been people, one standing, one lying down, the ghosts of old miners or Athabascan Indians. Who could know what history recollected itself in the dark absence of the trail at night?

Just ahead of us, the trail forked. I tried to remember which one to take. "I'll walk on up ahead and see where this one goes," Bill said. Leaning on my ski poles on overflow ice, I watched his yellow windbreaker dissolve. Then it was quiet enough to hear the wind high in the treetops. I listened for the squeak of poles pressing into hardpack. Nothing. What if he just disappeared? Just as fear began, I heard Bill's footsteps coming back, creaking on the dry snow. "The two trails join just a little ways up here," he called.

I hadn't waxed my skis or put on climbing skins, thinking the snow cold enough to provide the necessary friction. When I climbed the short hill to Bill, my skis slipped. I splayed my ski tips and herringboned. But the trail was too narrow, and I floundered, the tips sinking into deep snow while the backs slid sideways on the hardpack, my sled pulling me backwards and off balance. Bill stuck his poles into the snow, walked down and took the weight of the sled. I turned sideways to the slope and sidestepped to the top.

We paused there. "I remember the first time I entered a cross-country ski race," Bill said. We passed my water bottle back and forth. "I didn't have any ski boots. So I just took hiking boots and screwed them into a pair of wooden skis." Bill demonstrated the movements of turning the screws with his hands. "The race started at the top of a steep hill, and when I took off, I was just striding as fast as I could, and the screws came loose, and my skis went flying out from under me and shot down the hill. I just had to turn around and walk back with the screws sticking out of my boots like cleats." Bill hunched over his ski poles, wheezing open-mouthed with laughter.

Another mile up the trail I heard Bill stop behind me. I looked back and saw him cup one hand across his mouth, heard a sharp, short spray of aerosol. Bill's eyes were closed. I could tell he was holding his breath deep in his lungs. In the span of seconds it took for me to ask, "Bill, what is that?" a dozen scenarios passed through my mind. Bill having a heart attack there in the cold, hypothermia, asthma. How could I forgive myself if something happened to him? He was, after all, an old man.

"It's just a little something my doctor gave me to take if I get short of breath. I'm fine; I was just feeling a little tired. I feel great now. Good as new." Bill danced a few steps on the snow and grinned at me.

Wind shouldered our backs as we broke with the trees and slid onto the rumpled ice at the creek's mouth. To our right, stumps of Fred's woodlot splashed shadows on the snow. Across the moon's face, which had crested the snaggled treetops, clouds sheared. The lake spread like a plain, the mountains rising up steeply from its edges, snow billows rolling across the moonlit surface like tumbleweeds.

The trail was narrow so Bill and I stepped onto the lake single file. Although the weight of snow machines and boots had packed down a slab of snow, around us black patches shone like shards imbedded in wax, and I stopped, fear pin-prickling my armpits. I cast at the black with my ski poles and struck ice. Still, I had to remind myself as I skied along, that white smoke sheeting across the lake was snow and not steam rising from open water. Fifty yards out onto the lake, I stopped to wait for Bill. Our breath glittered as the wind took it.

Finally we neared the small peninsula poking out into the middle of the lake where Fred's cabin squatted. Pushing my skis forward, I spread my arms so gusts could catch the cloth of my windbreaker. I tacked into the wind, teetering on snow and ice across the lake.

Fred was asleep, buried under a pile of wool blankets, when we arrived at the cabin. The kerosene lamp burned low and yellow in the kitchen, and as we crept upstairs, its halo slid up the wall beside us. Even the massive barrel stoves couldn't pump enough heat to warm the second story of Fred's house, and we draped blankets and quilts over our sleeping bags and wore our hats to bed.

In the morning I stayed upstairs after Bill went down to breakfast. Friends had arrived after us in the night, and their laughs and chatter drifted up the stairwell. From the window, I looked at places on the lake whisked smooth by wind, imagining their steely hardness against my face. Between the lake and me, wind rattled the frosted panes, and a birch tree's branches scraped the glass like something wanting to be let in from the cold.

I'd wanted to feel spring at Carlson Lake. I'd wanted to listen to Fred's and Bill's stories. I'd wanted to take ski trips across lakes that beaded through the mountains. I'd pictured myself doing these things. But despite the blazing blue sky, the thermometer outside Fred's kitchen window read minus fifteen. I dreaded even the short walk to the tilted outhouse, with its plastic seat so cold I'd have to sit on my hands. A bank of depression settled inside my head, along my arms, in my gut. For me, it's a bodily experience, the way a drop in barometric pressure gives a friend of mine headaches, makes him sleep all day. No matter how many miles I ski into the wilderness, I can never leave this weather behind.

After a massive breakfast—sourdough pancakes with tinned margarine and corn syrup, sausage, canned bacon and fruit cocktail, hot chocolate, then butter cookies and a big sack of saltwater taffy circling the table in a continuous gyre—I came back upstairs and spent the rest of the morning writing in my diary. But something about the lake dragged my eyes compulsively toward it. Gauze blew across the ice. If I were brave, I thought, I'd skate on it.

Finally, I threw down my pencil and resolved to go out onto the ice, despite the cold and wind and my own lethargy, despite Fred's statement: "It's days like these that look best from behind glass."

I layered on all of my warmest clothes, wind pants and jacket, and pushed open the ten-inch-thick door. When I snapped my ski bindings into place with bare hands, my fingertips crimped, and I blew on them before stuffing them back in my mitts. I careened down the steep trail and strode windward on the lake toward large swaths of exposed ice. Even in the sun's brightness, the wind pinched my face until it felt taut and breakable. I pulled my neck gaiter over my nose and bent forward. There was no glide on the skis, just as much shuffling as I could muscle against the wind and the snow's chiseled surface—"sastrugi snow," it's called.

Ahead of me, exposed ice reflected a dark, oily shine. I carefully slid my skis onto it. Even though I knew that it was several feet thick, I tapped ahead of me with my ski poles, then stopped and stared down at the ice.

It was black with the hint of green, the color of air between trees in a spruce forest at dusk. Pinned like long sheets of just-developed photographs drying on a line, cracks hung below the surface. They formed a lattice, like spokes of glass wheels all interconnected. Someone might have explained their geometry as lines of fracture, stresses, and yet they were something other. They could have clasped to the very bottom of the lake, I thought, until I saw the gray shapes of fish—grayling—swaying five feet below my feet.

I unclipped my skis and shot them back. They clattered, moving unbelievably fast, and within seconds of scraping the snow's surface, stopped. I knelt on my hands and knees and crawled over the ice. I took off my mittens and pressed my fingers against its rippled surface. From the shaded side, the cracks looked like streamers of wet tissue paper or dried fish skin. Between cracks, rings of bubbles like splattered cream ascended in a series of mirror images.

I looked up and there was Bill, his arms swinging. He sauntered toward me in his yellow windbreaker and mukluks, his face exposed. I walked over to meet him. His eyes watered, the cold intensifying the creases in his reddened skin.

"Bill, look at this ice," I cried.

"My," he said, "isn't this something?" He shuffled over the ice staring down at it.

I lay down flat. The lake was smooth and cold as a tooth against my face. Bill took pictures of the ice for his wife. He got in one or two shots, and then his camera battery froze. His square fingertips growing red and shiny, he fiddled with the camera awhile, then tucked it under his jacket, stuffed his hands into his moose-hide mitts, and turned into the wind and walked farther up the lake, looking down.

In the wintertime in coastal Alaska, one listens to the marine weather forecast, and the announcer gives gale warnings and storm warnings and conditions of sea ice. There's young ice and brash ice and new ice. There's pancake ice and shore-fast ice and pack ice and the fragile skin of ice that forms in bays where freshwater flows in. But here was ice so thick and so solid, there was a world caught in it, and Bill and I shuffled across its surface as if we were riding above the architecture of our own lives.

I imagined myself on a boat, and the water was liquid and not ice, and all those things that had anchored me to life were falling away, deeper into the water, trying to find bottom, and I couldn't go after them. A woman once took my palm and divined the meanings of the lines there, the head line, the heart line, the life line, the creases for marriages, the crosses and vertical slashes for children. She traced possible paths with her fingertip; each line veering off the main one suggested a different outcome. Only one led to long life and the shape of a fish, some kind of spiritual awakening.

I turned from watching Bill and ran in the other direction across the ice and slid as far as I could. I ran and walked and skidded over to the edge where the lake struck land. First the world below me turned a greener shade of black. Then it turned a milky shade of green. Finally, the sandy bottom rose until I saw vegetation caught mid-sway in the ice, which was welded to the bottom like a barnacle to a rock. Grass and horsetail and the decayed fronds of water plants were caught in the act of buoyancy, frozen in their live forms.

On the other side of the lake, two years before, I'd seen the bones of a caribou killed by wolves. Did they balk when they came upon bare ice, as I did when I saw that first obsidian shard in the darkness?

Bill came toward me with the wind at his back. I ran and slid to meet him. My face was ready to crack from the cold; even the wind-dried surfaces of my eyes felt tight. "I'm ready to go back," Bill said. "How about you? I'd like to come here again after I warm up my camera and take some more pictures for Nancy. This ice—it's so strange." He shook his head, a grave look on his face, almost concern, as though we'd stumbled upon something dark and full of significance, like a corpse. I jogged over and clipped into my skis, and Bill and I shuffled across the slabbed snow to shore, climbed the hill up to the cabin.

That night Fred put a thermometer down at the lake so we could see the difference that the slight elevation of the house made. We made ice cream with creek ice. All night, the lake cracked and groaned. Then I knew that what looked so solid and of a piece—that whole frozen lake—cried out from the strain of holding itself together. I imagined it shattering, huge fissures opening like crevasses, like splittings in the earth at fault lines. Fred told us that during the great earthquake of 1964 the lake did shatter; the ice broke apart, and water spouts shot into the air like geysers.

In Fairbanks the previous winter, my friend, a fellow marine biology graduate student, who, like me, suffered from depression, had killed herself. For weeks afterward, I hadn't been able to sleep, so I got a prescription for tranquilizers and carried the bottle everywhere with me, as a kind of weird talisman. An out. Not able to convince her to go on living, I'd failed in the gravest way, and I didn't know how to react.

I scattered the pills onto asphalt and rode my bike over and over them after her father called to tell me the autopsy results, that she'd died of blood loss, not hypothermia, as we'd originally thought. She'd slit her wrists. The coroner had begged her father not to view the body. "If it were my daughter," he'd told him, "I wouldn't want to see her that way." I'd seen her footprints in the snow, imagined her lying down under the full moon, passing quietly and poetically away. In her father's broken voice, I heard the unpoetic truth, heard his spirit split through its center, though it didn't break, as Bill's didn't break when he lost Billy.

As a naturalist, I study and seek out nature's darkness—the partially eaten, eviscerated moose carcass; severed porpoise heads, cleanly skinned by killer whales, bobbing above the still-inflated lungs. Nature is perfect to me, in its means and its ends, indifferent and, like the lake, terrible and beautiful in its indifference. Yet, like Job, I turn my back on nature and rage at my childhood God when human tragedies strike.

And though he is an elder, I can't even ask Bill why. Watching Bill tells me that there are losses out there I haven't begun to imagine, but the world is a good and a sad place to be seventy-six, a place he's chosen to remain. His own depression has forced him, literally, to choose, again and again. Once I spotted a flash of color, a movement out of my upstairs window. It was early spring, and Bill was out walking on the trail through a stand of slender poplar saplings. He was bent slightly at the waist, bare-headed and -handed and wearing his homemade mukluks. His feet looked huge. Beneath the mukluks I knew those feet were horny and callused, flat and broad from going barefoot all summer long. His walk was steady and purposeful, and mostly it was moving toward something, something worth hurrying forward into.

After we returned to Fairbanks, Bill bought a pulp hook for Fred, who cuts all of his firewood, eight cords a year, by hand. He carved our names in its handle and sent it as a gift for our stay at Carlson Lake.

Bill and I talk about Carlson Lake often, and we wonder at its hold on us. We talk about it as if it were not a place, but something that happened.

When I see the lake in memory, we are walking on the broad, black, swept-clear ovals, and I am watching Bill as he walks windward, and I am considering the cold gleanings of ice, beautiful and strong, an intricate matrix almost but not quite breaking.

Somewhere Down That Crazy River

for S——

And night is a river bridging
the speaking and the listening banks,

a fortress, undefended and inviolate.

There's nothing that won't fit under it:
fountains clogged with mud and leaves,
the houses of my childhood.

—Li-Young Lee, from "Pillow"

The past is a place—a room, a house, an atmosphere, a lens, a river. In memory, we walk its long corridor; through our feet we feel its smooth oak planks. We wander along its banks, stubbing our toes on sharp rocks. The killer whales I study inhabit secret rooms beneath the ocean's surface. For five summers, I lived on Squire Island—a kind of secret room—with Molly Lou. Our house was a canvas wall tent nailed to a plywood platform, warmed by a corroded sheet-metal stove. We chopped beach logs for heat, their salt coloring the flames blue and green. Our research required that we, on the *Whale 1*, rove the area's labyrinth of islands and passages for months, searching for and following whales. When we couldn't find them, we found, usually not people—Prince William Sound is roadless and remote—but evidence of a more-populated past, abandoned towns, mines, herring salteries, cabins, shipwrecks. We found bears, deer, hidden ponds, and berry thickets. We encountered silence, when, after weeks alone, thinking and speaking blurred like the outlines of islands during storms. Weather enveloped us, even our minds. The absence of what we were looking for—the whales—swam around in our silence.

The Sound—a deeply indented cupping in Alaska's gulf coast—is a rain forest; rain falls, sometimes continuously, for a month straight. Even during good stretches of weather, every few days, a storm blows through. Once, looking for shelter forty miles away from our field camp, Molly Lou and I navigated our boat among the submerged rocks of Roberts Bay, on McPhearson Island. We tied up to an old dock. On shore, smoke wisped from the chimney of a ramshackle house, its front porch leaning toward the beach.

Now I imagine what I couldn't imagine then, the faces of Dora and George, moonlike, peering at us through rain-glazed windows, and us, the inevitable, almost expected arrival, making landfall on their shore.

Dora and George lived on McPhearson Island for five years, caretaking a small oyster farm. Besides Roberts Bay, McPhearson Island, thirty miles by boat from the nearest town, is uninhabited. Spruce forests and boggy meadows called muskegs ring the house. On the mountaintops, centuries-old hemlocks grow waist high, thick around as arms and legs. Their branches stream eastward away from their trunks like blown hair.

From the small-paned windows of a house clad in weathered planks and a tin roof, Dora and George kept watch on the weather, waited for boats bearing visitors and mail. Out-buildings—a guest cabin, a generator shed, and a three-story boat barn decorated with stained glass windows—flanked the house.

During that first visit to the island, in 1989, Molly Lou and I followed Dora's stocky frame as she led us on a tour of the barn. Like a bodybuilder, she walked stiffly, a little bowlegged in tight jeans but she stood barely five-two. She had green eyes and thick, waist-length hair that she wound into a bun and tucked into a white crocheted pouch. Her words burbled like a snow-melt creek, describing the Norwegian family that had lived on McPhearson Island for almost fifty years, running a boat-building and fishing operation. At one time, they skidded fishing craft up the beach and into the barn by hand. I imagined those boats inching up the slope, metal creaking, muscled arms straining as they cranked the winches.

Remains—wood scraps, broken-down generators, rusting barrels, sawhorses, coils of chain—littered the barn's floor. Hand tools hung from nails. In one corner, Dora had heaped garbage she predicted would never be hauled away. Even the air was old, thick with sawdust, wood rot, and grease, as if, in corners, it hadn't ever been disturbed.

On the second floor, Dora had arranged the Norwegians' household flotsam on shelves. She wove us through aisles like a museum docent, picking up artifacts: toasters; velvet paint-by-number paintings; a ceramic statue of a man and woman, nude and intertwined; cans of paint thinner, epoxy, and glue; lampshades. She'd spent hours up there during storms, she told us, ordering the jumble of lives she'd never known.

On the third floor, the old net-loft, I stepped tentatively on loose boards. When I touched mounds of cotton and nylon net slung over roof beams, dust billowed up. Even now, I imagine them waiting patiently for wetness, for rough hands, fish and salt, for life to return, the barn itself a net through which lives passed briefly, caught in the webbing like gilled fish.

Dora and George skirted those previous lives. In the main house, everything had its place. The kettle sat at a particular spot at the back of the stove, or the water would boil off. If I offered to bake something, Dora scurried anxiously behind me. A misplaced measuring cup was an invasion.

In the living room, Dora had arranged the house's hundreds of books and magazines—classics, Harlequin romances, cookbooks, nature guides, thrillers, mechanical manuals, *National Geographics*, Erma Bombeck paperbacks—on shelves, alphabetically and by subject. She balanced the oyster farm's accounts at a table near the window. I imagined a bear passing by as she bent her head over figures.

That night, Dora and George, knowing we'd been bathing in ponds and streams for a month, lit the wood-fired water heater and filled the claw-foot tub upstairs. It was an elaborate procedure, so Molly Lou and I shared a bath. Like children, we leaned against the porcelain surfaces of either end of the tub, our chins resting on our knees, our hair coiled in buns, the loose strands curling at our necks.

When Molly Lou and I couldn't work with the killer whales because of bad weather, we spent our time searching for edible plants, telling stories, sleeping under trees in our raingear, swimming, tying the boat to tiny islands and crawling through the brush on our hands and knees, filming the understory with a video camera for a movie about our lives in the Sound. One night, we filmed ourselves talking in the dark on the *Whale 1*. Our visit to McPhearson Island was a break in our habits, our isolation. "We" expanded from two to four; it gave us back another version of ourselves.

In the tub, we peered around the bathroom, dimly lit by two oil lamps. Dusty soapstone chunks George used for carving spilled out of the unused shower stall. In the small window above the tub, Dora had arranged feathers, candles, shells, and bones.

Later that night, the four of us, full from dinner, sat up late at the wooden table by the window. Staring hard past our reflections, which wavered like salmon in the glass, I tried to make out the star of the *Whale 1*'s anchor light.

From Valdez, sixty miles away, the radio on top of the refrigerator picked up Annie's Saturday night show. Her voice, gravelly from too many cigarettes, too much beer, could have been traveling to us from Asia for the true distance between us. We sat quietly after hours of talk.

George announced suddenly that he wanted us to hear his favorite song. Reaching across the table for the microphone, he hailed the marine operator on the VHF. There was no telephone on McPhearson Island; the VHF was Dora and George's only link to the outside world.

"Johnstone Point, this is the *Royal*, how do you copy?"

The operator put him through, and Annie's voice scratched across the line. Tilting his chair back on two legs, his stringy, red ponytail swaying, George grinned, his eyes fixed in their perpetual squint, a toothpick stuck between his small, brown teeth. His t-shirt stretched across his belly.

"This is King George from McPhearson Island, how're you doing, Annie?"

She laughed, playing along: "How're *you* doin?"

Dora smirked, rolled her eyes at George's cockiness, his trying to be suave. He'd never met Annie. In fact, he'd only been to town once in the three years he'd lived on the island. But she was their Saturday night friend, as much as anyone.

Near midnight, she played his song, Robbie Robertson's "Somewhere down the Crazy River." George told us that when they'd first started seeing each other, he'd given Dora a tape with that song on it. "Anyone wants to be my lady has to understand this song," he'd told her.

Robbie Robertson's voice, slow and sultry, described a man wandering in the dark, meeting a mysterious woman outside Nick's Cafe. She told him, *That voodoo stuff don't do nothin' for me.* Dora and George sang along, staring out the window, catching one another's glances, laughing quietly, *Yea, that's when time stood still.* George shook his head, spit into his cup, sang, *This is sure stirrin' up some ghosts for me.*

The world shrank down to the circle of lamplight, the music, the cups of tea growing cold on the waxy tablecloth. It was like watching two people make love. I looked away to the window, where the glass reflected our images back, flaming, so I couldn't pass through.

When the song ended, I was relieved. George got up to put more wood in the firebox, poured hot water.

Hours later, we rowed out to our boat. Phosphorescence, a green, glittering aurora, swirled around the oar tips. Molly Lou pulled the oars, and I watched from the bow. Between our skiff and the shore, the dark lengthened. The moon rose slowly, a ship's lamp above the island's eastern ridge. Dora and George waved at us from the beach, calling out above the warm wind, black figures against the lit house. As we bumped against the anchored boat, I saw the bead from Dora's flashlight bob as she walked the path to the generator shed. Later, barefoot, I stepped outside onto the wet deck to brush my teeth. With the moon behind a cloud, everything was dark on the island except the bedroom's pale light, shimmering on the water to our boat, tying us to them, like a line drawn in the mind between stars.

In our bunks, Molly Lou and I talked about Dora and George, how we'd entered their world so seamlessly. Molly Lou was twenty-one, I was twenty-seven, and they were in their mid-forties. While we were preparing to return to college and graduate school, they were preparing for their first winter on the island. They unnerved us a little with their openness. They were both alcoholics, "in recovery," they'd said. Dora had been dry for seven years, George for just eighteen months. They'd showed us dusty, unopened liquor bottles on the pantry shelves, left behind by the Norwegians. Instead of throwing them out, they planned to leave them there through the winter as a test of their will, their faith in God.

On ancient nautical charts, cartographers depicted the margins of the known world as an edge guarded by fantastic, bug-eyed sea creatures with mermaid tails and maniacal mouths. Such places are glimpsed from a ship or train car—one lighted window in dark countryside, an abandoned lot, sheets flapping from a fire-escape clothesline, someone stumbling, drunk, along the street—and recognized by an ache inside. Memory's margins—what we want to forget or conceal—set us apart, we think, from everyone else. When we recognize this wilderness in another, the intimacy frightens us; we sense something about to occur. The Eyak people, native to the rain forest, the Copper River area of Prince William Sound, named rain "something is happening." On the margin, something is *always* happening or about to happen.

McPhearson Island was such a place. And so is the Sound itself, a margin where polite conventions fall away, where friends bathe together, shit in the woods, pee in buckets, expose secret wounds, lose track of time, blabber frantically to the first stranger they see. Characters thrive here. While I'd learned to inhabit remote spaces since coming north to Alaska, Molly Lou had grown up in them. When she was a little girl, her family drove to Alaska from California. Her parents built a homestead on a sea-facing bluff. As a child, Molly Lou read by oil lamp light, bundled herself in coat, hat, and mittens to go to the outhouse in winter, sewed her own clothes, stitched purses out of alder leaves and spruce needles, dodged moose on her way to school. I grew up on a different kind of fringe of American culture, in an immigrant family. My parents were refugees from World War II. My male relatives, who fought in the Latvian division of the German Army, drank hard, ate slabs of head cheese on black bread, read Latvian newspapers, hated Russians, called themselves "foreigners." For me and for Molly Lou, the margin was a familiar place.

So we could visit Dora and George more often, we stashed fuel in their boat barn. They helped us roll the fifty-five-gallon barrels from the barn to the water's edge, where we'd beached the boat for fueling. When Dora leaned across a barrel, she curled her fingers beneath the lip and braced herself. "One, two, three," she called out, and we heaved the barrel upright. I felt almost no resistance.

"Dora, let me take some of the weight; I don't want you to hurt your back," I pleaded.

"Look at my arm," she said, pulling up her sleeve and placing her forearm next to mine. It was as thick as my calf. "I've gotten strong out here. You should'a seen the looks on the barge guys' faces when I threw fifty-pound fish food boxes at them. They just stood there with their mouths open." Dora brushed her hands together and turned back up the beach for the next barrel.

Her arms reminded me of a man I saw once. Driving north between Anchorage and Fairbanks, the road a slab of packed snow, in the distance I'd seen a shape humping along the right-hand shoulder and at first thought it was a child pushing himself along on a sled. As I'd passed, I'd realized it was a man walking on his hands. He had no legs. He wore a green canvas knapsack on his back and every few feet, he set himself down to rest.

In September, Molly Lou flew east to college, and I drove four hundred miles north to Fairbanks for graduate school. I wondered often about Dora and George. I pictured those dusty bottles, the house buffeted by the eighteen-hour night and winter storms. They didn't even have a decent boat to get around with, just a flat-bottomed, raft-like creation powered by a salt-encrusted outboard with no cowling, requiring mysteriously violent maneuvers to start. But gradually, I immersed myself in my other life. In that life, mechanics fixed my car when it broke down while I pored over books and journals in libraries. On computers, I analyzed data Molly Lou and I had collected over the summer. I forgot about Dora and George.

The next spring, Molly Lou and I returned to McPhearson Island. As we docked, I told myself not to be surprised if Dora and George were gone, but there they came, puttering up to meet us on the raft, beaming huge smiles. "How was your winter?" Molly Lou called across the water separating our boats.

Dora shouted back, "It was the best six months of our lives—we got married!"

At the kitchen table, George showed us a soapstone carving made by his father and one he'd been working on himself that winter. "And this," he said, pulling a pendant from a jewelry box, "was given to me by a special lady." He cradled the amber chunk in his palm and smiled. Dora snorted, shook her head.

Through the afternoon, George told us his story. Violence ran through it like a coal seam. He grew up in a town with a name that seemed to define his life: Concrete. Once his father, drunk and brandishing a shotgun, broke into the shed George slept in. At eighteen, George was drafted, went to Vietnam. After his tour, he re-enlisted twice. When he returned to Concrete, George took his father's place at the drinker's bar. He worked on logging operations, then fought forest fires. With his suspendered dungarees and work boots, he still looked like a logger, but at some point, he'd read Edward Abbey's *The Monkey Wrench Gang*, and now he wore a faded Earth First t-shirt under his wool jacket.

I asked them if they'd been lonely over the winter.

"Sure," said Dora. "Once all the boats left in the fall, it got pretty quiet out here. But Ellen and I didn't worry about talking on the radio then; fishing was over for the season." Ellen and Steve, an older couple, lived in a cabin in another part of the Sound. They set-netted commercially for salmon in the summer. "We got on channel eleven every night and yakked for hours. We traded recipes and griped about the men." She grinned at George. He leaned back, crossed his arms over his chest and smirked.

"How's Steve doing?" I asked. They'd told us before that he drank, and that, given their history, it was hard to be around him, though they craved the company.

"Well, we had a scare with him this winter," said George. "Ellen went into town in December to get her teeth fixed. Steve went on a drunk. One morning we heard a call for us on the radio. We were still in bed."

Dora leaned forward. "George practically fell down the stairs trying to get the radio," she said. "He always hears the radio, no matter how asleep he is. Well, it was Steve. He was babbling away, telling George that he had a loaded gun to his head, that he was about to blow his brains out. When I heard that, I ran downstairs and grabbed the mike out of his hand. I knew I was the one to talk him down."

I thought of my own winter, how I'd analyzed data, presented my results at a scientific conference, passed my comprehensive exams, organized protests against wolf control. My life had seemed purposeful. Then, in January, my friend had spiraled into a suicidal depression. My old life stopped. I tried to talk her down, but on a forty-below-zero night in February, she bought a knife and drove north. She stopped at a diner, bought a cup of coffee and turned her truck back toward Fairbanks, then turned right on a single-lane road. She dissolved a bottle of antidepressants into the coffee, slit her wrists, and walked off into the wilderness.

What keeps order in this world? The scientist with her number columns? Dora and George organizing the barn, watching weather and currents and what drifts up on shore, listening to the radio? Their stories sleep as mute as stones through the long months, and then, like geodes flung to the ground, crack open to reveal their crystalline insides.

I looked around the brightly lit kitchen. Imbedded in seams between floorboards worn smooth from fifty years of sock-footed shuffling was dirt. Upon the walls, events left tracings, like breath on glass, like dust coating the liquor bottles on the pantry's bottom shelf.

After dinner that night, Molly Lou and I followed Dora up the narrow stairwell to the guest bedroom. It had been the children's room when the Norwegians had lived there. As she climbed the steps, her flashlight beam scissored over the faded wallpaper.

Molly Lou and I wandered the room, picking up books and looking at the odd assortment of furniture and half-finished projects lying around. Dora's flute rested on the corner of a bed. Two easy chairs faced one another with a low table between them. Dora and George played backgammon there, or monopoly, or cards. She held up her sewing projects, which she'd draped over another chair. I brushed my fingers over a half-finished rag rug she'd worked on the previous winter.

Later that night, Molly Lou and I crawled into the sagging brass beds by the window facing the bay. The sheets felt like baby powder against my skin. We called out good-nights to Dora and George, their light shining under the bathroom door, and I heard Dora moving about. "Sweet dreams, Eva," said Molly Lou quietly.

"Sweet dreams," I whispered, but I couldn't sleep. I stared at the window thinking of the children who'd slept in that bed, at what imaginary worlds they might have made of a night like this, of the wind's drift, of the creaking and settling of the house. I've always been afraid of the dark, more so after my friend committed suicide, but that night, I stared out the window at the bay and imagined all of us as children, safe on the island.

The next day, heading back to our camp, I confessed, "You know, sometimes I have a hard time with things George says. Like the way he says, 'Don't fuck with me.'" We laughed at the imitation, but she knew what I meant. Extricating ourselves from them was hard. They'd make us logger's breakfasts, stacks of pancakes, canned bacon, cowboy coffee. Then the stories would begin. We shut down the engine in the open water south of McPhearson Island. The passage lay glazed and slow in light filtered through high clouds.

"I can't help thinking it's just a cover, that George doesn't mean it. It's his image; without it he'd be too fragile," she said. But it was more than that. In town, in a bar, I could imagine writing George off as a redneck asshole. He boasted that, in the war, he'd shot people, and back from war, he'd "busted heads." He'd perched a shotgun conspicuously above the kitchen door. He wanted people to be afraid of him, but we weren't. They revealed too much to judge them easily. Dora told us that she watched George stalk a river otter once on the lagoon bank. He cocked his rifle against his cheek, stared through the sight but couldn't fire.

Molly Lou and I were quiet. We didn't always answer each other's questions right away. Sometimes they hovered in the air around us for days. Sometimes I couldn't differentiate thoughts from words; it didn't seem to matter.

"I wonder what George and Dora are doing right now," Molly Lou said.

The air smelled pelagic, like salt and kelp and swells. Not even a distant boat engine hummed. A jaeger tumbled through the air near us, chasing a kittiwake, their cries cutting the stillness. The kittiwake laughed, winged up and over the boat, and flapped away. The jaeger dropped to the water and dipped its bill into the colored snakes of its reflection.

On our next visit, the four of us in the kitchen, I watched Dora trace the tablecloth's pattern with her fingers. "When I have bad days, that's when hugs come in," she said. "George's good at that. Lots of hugs. Sometimes, when I'm just about to go crazy, he walks in here and holds me until I calm down. He's a good hugger." George stood, grabbed her hand, and pulled her up. They danced, bodies locked.

"Keep going," said Molly Lou. "It's great."

I related to Dora's bad days. She kept an assortment of herbal remedies in the glassed-in cupboards where the dishes were stacked: valerian root, chamomile tea, ginseng, a homeopathic remedy she called "calms." She told us she couldn't deal with too much stimulation. Crowds of visitors, rambunctious children, strangers, men, all could unhinge her. She ordered her life minutely against chaos.

I recognized that impulse. As children, while my parents worked, for fun my sister and I moved the living room furniture outside so we could sponge the walls and clean the carpet. I envied girls whose mothers were obsessively tidy. As an adult, I'm distracted from work by an unwashed dish, mislaid papers, a clump of dog hair on the carpet. I clean the kitchen *before* I cook. At camp, during foul weather, Molly Lou and I scrubbed and rearranged. A visitor, another biologist, once commented, "What does it matter? This is a *field* camp!"

On McPhearson Island, the bay's rock-encrusted entrance held danger and chaos away. As long as Dora kept sorting and organizing, someplace was safe.

July. *The heart of summer*, Molly Lou calls it. In my memory, we're heading to the island for fuel, our third summer of knowing Dora and George. "Let's make them a special dinner," I say.

A fish-buying tender, the *Hana-Cove*, floats in the bay, four young men on deck jigging for halibut, cigarettes dangling from their mouths. They shift poles to one hand and wave as we pass. On the beach, Dora and George clean oyster trays with a pressure washer. They're in full raingear, despite the hot sun.

When he sees us, George strides down to the water's edge and reels in the *Royal*'s tow line. Molly Lou and I stuff our day-packs with clothes and journals, and I search through our food stash, looking for a treat to bring them. When I lift my head from packing, the raft is gliding in toward us. I jump out of the cabin, throw a line, and pull George in as he kills the outboard. Molly Lou tosses our packs across to him, and we climb over.

"Good to see ya," George says as we hug. "Had a feeling you'd come, heard you on the radio. We've been working our asses off. Just got another shipment of 400,000 oysters from the boss. Dora'll sure be glad to see you; she could use a break."

George sits, legs splayed, on an overturned bucket and yanks the pull cord. The engine sputters to life. As we rumble shoreward, I secretly search his face. His skin seems less gray and his eyes clearer each time I see him, and I wonder if it's just my growing familiarity, affection instilling its own kind of beauty. But I want to believe it's the island.

"What's the *Hana-Cove* doing here?" I ask.

"They pulled in this afternoon to get some oysters. Fishing closed down for a few days, so they're taking it easy. We traded them for a couple T-bone steaks and a turkey. Been a long time since we had meat." He tells us that the *Hana-Cove* was a drug smuggling boat in Chile before converting to a fish tender. Now the Falls Bay boys run it. They have a mean reputation, he says. This is George's kind of story. It gives him something to protect us from.

As the raft scrapes shore, Molly Lou and I leap off. Brutus, their dog, a long-legged, brindled boxer with a pushed-in face and a friendly disposition, snuffles around us. They once told us he used to belong to Gary Larsen, the comic strip writer. Dora calls him "Muffin."

"Why don't you go on and visit Dora. I'll pull up the shrimp pots."

Dora works under the tarp. There aren't enough screens on which to plant the new oysters, so she's consolidating, cleaning mussels and starfish off the ones they already have. We drop our packs and help.

"I used to like starfish," she says, shaking her head. "Not anymore. You wouldn't believe how fast they build up on the oysters. I don't think this is the best place to grow these things. What is the boss thinking, sending us all these oysters and no one to help out? I told George I was ready to quit."

This is familiar. Dora and George often threaten to leave, after they've saved up enough money. Dora does most of the oyster work, but only George gets paid. George mainly fishes, hunts, and tinkers with engines and generators.

"The boss is clueless," I concur, shucking mussels.

"He always comes here when it's that time of the month." Her reddened fingers fling shells. "It never fails. It takes everything I have to just bite my tongue until he leaves, even though I want to scream. And the wife, she comes here and starts cleaning, scrubbing the walls and moving stuff around so I can't find it. As if I can't keep house."

The boss doesn't understand what a good thing he has in Dora and George, I think. But I know Molly Lou and I see them through a prism, and turned a millimeter, the picture, to someone else, would be different.

"Dora, Molly Lou and I want to make you a special dinner, what do you think? Anything you want, we'll make."

Dora looks up from her work and smiles. "Hey, girls, that sounds great. We just got some steaks from the *Hana-Cove*; I'll go out there and get a couple more." I hesitate. Neither Molly Lou nor I have eaten meat in years. Molly Lou raises her eyebrows, then shrugs. What the hell, I think, this is special.

As George eases the raft toward the mooring buoy—he's returned from pulling shrimp pots—Dora strips off her slime-covered rain jacket and ambles down the beach to reel him in. I watch her compact body leaning back with every pull on the line. She's wearing snug jeans, a sweatshirt with the sleeves bunched up around her elbows, a faded blue bandanna around her neck. After George gets out, she shoves the raft away from shore, scrambles aboard and jerks again and again on the pull cord until the engine lurches to life. She settles onto the bucket, leaning forward, one hand on the tiller, the other on her knee.

Later, Molly Lou and I bustle around the kitchen. Dora shows us where the flour is, in the bread room, and the cans of fruit and vegetables in the pantry, then goes out to finish work. Searching the pantry's bottom shelves, we wonder at grimy jars of jam and pickles floating in cloudy liquid, put up by a previous inhabitant. Nothing here is thrown away.

Not having heard a radio in weeks, we switch it on. Laughing, we dance to oldies around the kitchen. Molly Lou puts on an apron and mixes dough for peach tarts. She whips the batter, considers it, adds more flour. We use ingredients we normally shun: Crisco, white sugar, white flour. I fry up the steaks on the wood cookstove.

"Molly, I'm not sure I know what I'm doing here; I don't want to wreck these. We'll never hear the end of it from George."

Molly Lou scoops confidently into a tin can on the warming oven and drops grease into the pan. "I don't think you can go wrong with this stuff. Just don't overcook them. I'm sure George likes his steak raw. We should make gravy, too." We mix up cornbread, heat canned green beans with margarine, fry onions with the meat. Grease spatters, burning my arms.

"How's it goin', ladies?" Kicking off his gum boots, George hangs his wool jacket on a peg. "Smells good in here." He pours himself a cup of inky dregs from the pot on the stove before wandering into the living room.

When the steaks are done, I put them in the warming oven and set the table, picking out the least-chipped dishes from the cupboard. On the lower shelf, behind plates, I find four wine glasses.

"George, what should we have to drink?" I peer around the door. He's slouched on the couch reading a paperback.

"Oh, just the usual," he answers. I mix up a pitcher of Tang and pour it in the wine glasses, then go outside and pick salmonberry blossoms, daisies, and an iris growing near two huge spruce trees. A hammock stretches between them, and I climb in, looking at the light filtering through branches as the swaying gradually stops. This is Dora's favorite spot, where she comes when she's feeling overwhelmed. She writes poems here, some of which she's read to us.

I think of the nightmares Dora's told us about, men coming to the island while George is off hunting, men breaking into the house and raping her. She believes that her thoughts have power to make things happen or to hold off disaster. She believes she must keep her thoughts under control.

Once, in winter, George left on a two-day moose hunting trip, and a shrimper, the *Wolf Kill*, anchored in the bay. A crewman came to visit and stayed too late. Dora was nervous. He flirted with her, though she hinted that George was due back any minute. I don't know how she finally convinced him to go.

She keeps a loaded gun by her bed. The rifle over the kitchen door is loaded too. "As soon as you walk in the house, it's staring you in the face," George said.

"I've gotten better at being alone when George is away," Dora told us once. "You should have seen me at first. I was a basket case. Now, we don't like it, but Muffin and I get by."

Back inside, I pour water into a jar, arrange the flowers for the table. "Dinner's ready, George." He's fallen asleep on the couch. Molly Lou goes out the side door to call Dora.

George always sits at the end of the table facing the window. "Gosh, you guys, this is a feast!" Dora exclaims.

George laughs. "Let's see how well a couple of veggies can do with steaks."

"Can we hold hands for a minute?"

"Sure," says Dora. I reach for Molly Lou's hand, and George clasps my right. I keep my eyes open for a second, watch Molly Lou and Dora lower their heads. My heart pounds hard, as if I'm scared.

George speaks, his voice pitched higher then usual, "Lord, we thank you for the gifts we are about to receive, for this food and for these friends. Amen." When we open our eyes, they're glistening. We smile.

"I love you guys," I say, quickly.

"Us too. Now, let's eat." George grins and jabs his fork into a T-bone steak.

Later in the barn, I pump fuel into a jug by flashlight while Molly Lou carries the filled jugs to the water's edge. We want to get an early start in the morning. When I'm alone in the barn, the darkness curls around my shoulder like an arm. My hand clutches the pump handle, paralyzed at the top of a rotation. I peer around at humped shapes. When I hear Molly Lou's boots digging into the beach stones, I breathe.

"You wouldn't have liked George back in his drinking days," Dora said earlier that night, after we'd cleared the table. "When we first met, he was drunk all the time, getting in fights, not coming home. I wasn't drinking anymore, but I put up with it at first."

George broke in, "I came home one night, and she was pissed off. We got into a major fight, and Dora tried to slap me in the face. No one does that, and I was trashed."

"I could see it coming," said Dora. "*Oh, shit,* I thought. He just pulled his arm back, and I saw his fist coming at me. I tried to dodge it too late, and he got me right in the eye."

I didn't want to hear this story, wanted to rewind it.

"As soon as my fist connected, I knew I'd fucked up, but it was too late . . ."

Dora cut him off. "I told him to get out. It was over. I'd told myself no one would ever hit me again."

"That was the last time I had a drink," said George. "Took me awhile, but she finally took me back. That's the first and last time I ever hit a lady."

"It'd better be," said Dora, laughing. "I walked around for weeks with this ugly black eye."

That night, on the boat, Molly Lou and I lie on our bunks and write in our diaries. After we switch off the light, George and Dora's story rears up like a grotesque shadow on a wall. How much of their strength depends on the island? "Do you believe them?" I ask Molly Lou, speaking to the darkness. "Do you believe George won't drink again, that Dora won't let herself be hit?"

"Yes," she says.

I wake suddenly, my body tense. Wind spins us around the anchor, the boat's movement inscribing a half-moon on the water's surface, the actual moon casting a square of light onto the wall. The light slides across the wheel, seat, table and passes across my shoulder and face. Like a palm, it brushes my cheek.

On our next visit, in August, we picked berries. Dora scolded me when I squashed a strawberry under my foot. Someone had planted them years ago in the chin-high grass next to the barn. Dora never missed a berry, following behind Molly Lou and me, pointing out ones we'd overlooked. Behind the barn, wild blueberries grew in profusion. With the strawberries and Dora's wizardry, they'd make a huge pie, which we'd polish off later in a single sitting. George dropped handfuls of berries into the bucket, but mostly he ate and railed against the boss.

The boss kept sending more oysters, refusing to hire help, or to offer more pay. Brushing grass aside, I made up letters in my head to the boss. Was he blind to how irreplaceable Dora and George were? Was he crazy? I didn't care about his side of the story, saw only that Dora and George belonged on the island.

After each visit, Molly Lou and I wrote. We recounted stories to one another until they became glass paperweights we passed from hand to hand. We listened in on their radio calls, and they listened in on ours. We were loyal to each other, like children.

It was always hard to leave the island. That last time, after the pie was gone, George ran the raft over to Barnes Point to fish. Molly Lou and I packed up, then helped Dora stack dishes near the sink. She wouldn't let us wash them. She had her own method.

"You know," she sighed, sitting down at the table with a mug of tea, "George thinks he had it bad. But when he was over there, I was fighting my own Vietnam War." Something in her voice let us know we couldn't leave. Molly Lou and I sat back down.

She told us that the first time her father molested her, she was two years old. Holding her mug in both hands, her eyes fixed on the cool liquid inside, she described what he did to her straight out, and the images in my mind floated like spots, as if I'd looked straight into the sun. "I tried to tell my mother, but she didn't believe it. She called *me* a pervert. I was sixteen when I finally told him no. I had to get out of there."

"I know what you mean, Dora—," I began.

"I kind of figured that," she said, meeting my eyes.

How is it we all seem to recognize one another, I thought, like members of the same, strange tribe. The children who we were stand at a distance. Like images seen through binoculars backwards, their solemn figures seem impossibly far away. I long to smash my fist through the glass of their silence, as Dora had done with hers.

Dora swiped at the tears on her face with her palms. As she talked, the strands of her life coiled away in all directions from her childhood like a tangled mass of nylon line. Years of drinking and drugs, her children taken away from her. Checking herself into a treatment program. Going back to school to learn seafood culture. This strand curled out of the morass to the island, to the three of us around the table.

I imagined the house on the edge of a precipice, teetering. I understood her fear of disorder and my own. Until we're safe inside our own bodies, no place is safe, not even an island. A suddenly opened door might bring in a blast of wind and rain, a flash of lightning illuminate a figure standing there, and the house will tip.

She stood up and opened the firebox, stirred around embers, fed in spruce logs. Outside, the tide licked the beach. Our boat floated at the mooring. The *Royal* was out of sight, somewhere around the point. Brutus came to rub up against Dora, and she bent to scratch his wrinkled head. "Oh, hello there, Muffin, do ya wish someone would pay attention to you? Yeah, you're a love."

Later I asked Dora if there was something I could send her. "Just letters, just friendship, and maybe some chocolate sometimes and chamomile tea." She was bright again, sauntering down to the edge of the water with us, her arms swinging, sun hat shading her face. She lifted her hand to her forehead, searching for George.

Out on our boat, we skimmed the bay's glassy water, stopping at Barnes Point to say one last good-bye to George. He thanked us for staying with Dora, as if he knew why we'd taken so long. "You guys are good for her," he said. "She gets lonely for ladies to talk to." I smiled and hugged him.

Months later, in March, Dora called me in Fairbanks, her voice distant and broken on the radio connection. "We're leaving the island. I can't explain it over the radio. I'll write to you and let you know. We'll be out of here by the first of June. George is heading for town to look for work." I desperately wanted to know more, but she cut the call short because they were saving their cash.

The next time we spoke, I was out in another part of the Sound, and I called them on the radio. She was alone, George gone to Wrangell to work, Dora not even knowing what kind of job he had.

She was angry on the radio. "I'm just being the good little wife," she said, "following George. As usual he's left me with all of the cleaning and packing up." She laughed, as if to shrug it off.

"Come to Fairbanks," I told her, impulsively. "Stay with me."

"I'll try to come," she said. But she didn't, and I haven't heard from her or George since.

Every winter night in Fairbanks, as soon as it was dark, I'd pull down thermal window quilts to hold back the cold. My neighbor called my house "the cocoon." Saved things, revealed and hidden daily by the lifting and lowering of the shades, decorated the foot-and-a-half deep window sills: a glass ball found in the Sound; a photograph of Molly Lou and me in a fireweed meadow in Lucky Bay; an oyster shell; a picture of Dora and George on the porch at McPhearson Island. George's arm wraps around Dora. As usual, his look is wry and guarded, as if he's protecting something and at the same time trying hard not to grin. Dora's face is a luminous moon. She leans into George, but her broad smile seems to project her out of the picture. Behind them, the house harbors layers of time, in scars on the tabletop, places in the floor rubbed smooth, cracked glass, chipped plates, cracks in the plaster, water stains.

There's a kind of strength I've been looking for. Gained by accretion, day by day, it's knotted into the body like wounds in a lightning-struck tree, twisting a husk around history, a hard and brutal kind of beauty. It's Dora rolling out a pie crust with her two strong arms. It's the arms of a man who walks on his hands. It's George's eyes, guarding all they've seen in their squint. It's my own wrinkled-up hands and Molly Lou's finger, scarred where, in childhood, it was cut by the spinning wheel of an overturned bicycle.

It's said that the body is the soul's temple. When the body is violated, when memories of disaster inhabit the temple, scar tissue roughens the inner skin. Though we learn about survival, the scars remain, the way violence loses its power but is never undone. The past doesn't wash away like seaweed after winter storms. It inhabits the body, and the body is a house of secret rooms. A friend of mine built a sculpture once, shaped like the back of a humpback whale, with an arched opening viewers could crawl into. She painted it black. Inside it, the bones were visible, the structural elements that gave it strength. She called the sculpture *Memory House*. She built it in grief over the *Exxon Valdez* oil spill. In my memory house, I find Dora, her arms lifting candlesticks and lampshades onto shelves. She's paused to tell a story. All of our histories are held and come clear in that one healing act.

To the Reader Who, From the Eternal Present, Asks About the Oil Spill

Ask which beasts are
all gone and for how long
—if 1 oyster-
catcher garnet trills.

Still we highline, run her on the high
tide,
her beauty misleading—

her treasure.

We—full fathom
only little—:

—Molly Lou Freeman

This essay doesn't belong here. It's not chronological. It's from a now, a present tense that's gone the moment this sentence is finished. In this present tense, it's 11:30 a.m., the eleventh of August, 2006. Another rain squall envelops lower Knight Island Passage in Prince William Sound. Shreds of whiter cloud drift and morph as they brush the lower slopes of the islands, losing themselves in the forests, where blueberry bushes drip their sweat. But in this book, this present tense hasn't happened yet. It's a strategy, this placement, for dealing with a certain kind of event. Thus, this essay belongs here, where a silence gapes from the words that came before and the words that come after. Where a silence opens like a maw. Where a silence says, *speak to it.*

There's a silence in this book. It's the year 1989. Yes, I've told my stories to friends. Yes, I've testified at the appropriate hearings, where people tore their hair and cried. Holocaust survivor Elie Wiesel says, "Not to transmit an experience is to betray it." What duty do I have to this experience? What relationship to it that I could betray it? Are we connected, this experience and I? I refused to read about it. To go to slide shows after that first one when I saw the Pleaides Islands oiled not beyond my recognition. Wiesel asks, "Where was I to discover a fresh vocabulary, a primeval language? The language of night was not human, it was primitive, almost animal." Maybe I was waiting for that language to bubble up inside me, like crude oil sunk and hardened beneath a beach, oozing up when the sun baked it. Or maybe I was waiting for enough layers of silence to pile up to make an ominous creaking sound I couldn't ignore when I wrote my essays about place, about love. The silence creaking, floorboards bearing too much weight, a door unhinged, banging in wind gusts. Or maybe I was waiting for a structure to hang the details on, like slats for salmon in a drying shack. Or maybe I was just waiting for today. For the present.

"I see them and I write," Weisel says, meaning the dead. I see them in Long Channel, a narrow pass between two islands, a group of four transient killer whales, diving under the fishing boat *Humboldt*'s stretched seine. Rain pocks the surface. The skiff man, powering his engine with his left hand on the steering wheel, pushes one end of the seine net hard against the shoreline. His body, from a distance, looks like it's heaving a harpoon at the water. With his right hand he's heaving a metal cup on the end of a pole—a plunger—to scare fish trying to escape from the seine. He pauses with his plunging. The crew stacking corks and web into piles on the seine boat's deck turn their heads. We all watch to see what will happen, if the whales will entangle in the seine, feel relief when they surface several hundred yards down the channel. The fishermen resume their work. We follow the whales. I see those same four whales juxtaposed against the grounded oil tanker in a photograph in the Anchorage newspaper in the early days after the accident. The day is sunny. The hull of the tanker is black, but the killer whale fins are blacker. The last year we'll see those killer whale fins is 1989. The dives repeat, over and over. The number of imagined dives—like the number of carcasses of birds and animals—becomes too much to bear.

And the only balm, as always, is time. These days, hours go by out here without thinking about the oil spill. It's been a generation. For teenagers in the villages and towns, for my stepchildren, it's history, it's stories they hear. It's the story my stepdaughter Elli knows. During her first visit to the Sound, the beaches reeked of crude. She was an infant, and her parents worried about her inhaling toxic fumes. But the instant they heard, they could do nothing else but what anyone does in the face of a loved one's imminent death. They dropped the seines they were mending, the herring pounds they were building, bundled Elli in wool and pile, and headed for the Sound. For children of fishermen, it's the story of their parents flinging—not salmon, not herring—but oil out of the Sound in five-gallon pails. Those parents, those who haven't died or committed suicide, are still waiting for a monetary settlement from Exxon for lost fishing income after the spill. Six thousand. That's the number of people who've died with claims against Exxon. They are heavy stories for kids to bear, like the ones I heard my relatives tell over and over, how they escaped from Latvia during World War II on trains, on foot, on horseback, or how they fought on the lines, how they pushed the horrible memories down, how they waited for resolution all their lives.

We speak of it the way children of alcoholics or veterans speak of their damaged lives, how poison seeps through all of the strata, how every quirk and twitch might be attributed to the triggering incident. In Stockdale Harbor the other night, we anchored alongside the *Alexandra*, our friend Brad with a group of Japanese photographers, one of whom has come to the Sound for twenty years. Brad says he's been seeing eagle carcasses on beaches, and yes, come to think of it, we haven't seen as many eagles, have we? And after this observation is made, fear bubbles up inside me, marking a place, the way a stream of bubbles marks the spot where a sea otter went down, exhaling. It marks a place of disquiet, a possibility that doesn't jive with my intense desire for this place to be whole again, so the mind turns to more hopeful signs. Harbor seals seem to be coming back. But not herring. Not seabirds. Sea otters still ingest oil along northern Knight Island. After twenty years, how do you tease out the damage of one thing over another, lingering oil tangled up in changing ocean temperatures, increased small boat traffic, regime shifts, spewed oil tanker ballast water, natural cycles, some species thriving at others' expense? You just keep returning to that place, like a salmon, like a bird. You can't help it.

I look back and there's sickness and silence, look forward and there's fear. Only in this present tense is there solace. Is there—yes, I'll say it—love. It is a love story, in the end. The clouds have lifted over Point Helen. I can see Montague. I snap photos of the bay behind us, draped in fog tendrils and curtains and rags. It's the place we call the "inner sanctum" of the Sound. There is solace in the harbor seal that eyed us, floating by our anchored boat like a hippo, just the nostrils and eyes above the surface. Solace in the rain, which after a few hot, sunny summers, returns the Sound to its primeval, rain forest state, and we hope, keeps away all but those who love it for what it is—not for what they can take from it.

In relationship to the oil spill, the people still testify, repeat their stories, make their claims, mainly around kitchen tables, over breakfast at the Reluctant Fisherman or the Killer Whale Café in Cordova, in the galleys of boats weathering out storms. The beaches are still oiled. The Sound is changed. People's lives are changed. And there is cynicism, what comes up out of the deep, where injustice and betrayal are buried and unresolved, and like oil, harden into an asphalt only to ooze up when the temperature's right, and it stinks, like poison.

My friend taped a cartoon to the wall in the environmental program office in Chenega Village. Two people stand on a beach. One says, "When do you think Exxon will recover the remaining oil in the Sound? The other replies, "When the price of oil reaches $100 a gallon. Then they'll dig it up and sell it back to us."

I'm not a scientist when I talk about love, for me, being the primary mover in long-term research of killer whales. I'm a scientist when I say that oil inhaled through the blowhole can cause pneumonia. That fresh oil vapor inhaled through a blowhole can cause instant death. That crude oil ingested by way of oiled and lethargic harbor seals by killer whales can damage the liver. But who am I when I say that on that rock, there, the one that looks like a humpback whale's dorsal ridge, a seal once crouched? A killer whale swam back and forth, waiting for the tide to rise, to float its prey. Then something turned its head around, and the killer whale charged across the passage, calling very loudly, so I could hear it through the wooden floorboards of the orange inflatable raft. I dropped a hydrophone over. The lone whale was screaming. All over the Sound, I would drop my hydrophone and hear those cries: one whale, usually a male, looking for its relatives. Those calls have been part of the acoustic landscape of the Sound for probably thousands of years, but who would know it? Who would hear it?

I heard those calls today in lower Knight Island Passage. By the time I die, a silence will replace them, a silence so deep it leaves a vacuum the roar and scream of skiffs and seiners and freighters and tankers and tenders and yachts can never fill. There's a company out of Whittier that proclaims it can show you twenty-seven glaciers in a day from the observation deck of its high-speed catamaran. But don't be fooled. It's going so fast, you'll miss the emptiness, the silence, the spaces left by animals and birds that have died, not only because they were fouled by oil, but because they ate bits of decomposed plastic water bottles, because they were infected with a disease that thrives in higher water temperatures, because they were struck by high-speed boats. And you'll miss the other, bigger silence that holds all of those deaths, that holds what's left, that I imagine emanates out of the inner sanctum of the Sound as tendrils of fog, today, tomorrow, a thousand years ago, that is the essence of this place, the secret ingredient, the elixir, very much alive.

One of those lone screaming whales was Eyak, the first transient I recorded emitting sequences of long-distance-contact calls. He beached himself near Cordova, died under the shadow of Eyak Mountain. His skeleton now hangs from a museum in Cordova, reconstructed with care by schoolchildren and volunteers, and a photograph of this skeleton is on a bookmark that someone gave me, but it's another thing I can't—won't—look at. I've tried. The image marks another place of loss. I knew this whale! *So what?* you say. It was not a person. And what does it have to do with most people's lives, people who can't afford to heat their homes this winter, to drive their cars? My brother, a social activist, says we will never convince the poor to care about the environment until they can make a decent living, provide for their families. Why should someone in Cinncinnati care about the fate of twenty-two killer whales? My brother says that environmentalism is for the privileged. And besides, what do you mean, you, a scientist, *knew* that whale?

What is knowing? I admit I knew very little of its (*his*) life. He was AT1. *A* for Alaska. *T* for transient. The number one because it was the first named individual of a mammal-eating group, an assemblage of whales that traveled in smaller groups but periodically came together, like an extended family, with strong ties and a shared dialect. The AT1 group, with only twenty-two members in 1983, was probably genetically doomed even then because they didn't mate with other transients, but still, somehow, weirdly, until 1986, they had calves.

After years following him, learning that he spent a lot of time near Cordova, the ancestral home of the Eyak people, the male AT1 became Eyak. It became a *he.* He had a story. He had a unique personality. He had a distinct voice. Children in Cincinnati, I'm convinced, would understand. I once gave a whale talk to teenagers at a youth detention facility in Fairbanks, to kids, the facility counselor told me, who'd committed the most horrible crimes, rape, even murder. The kids sent me a group poster in thanks. One boy wrote, "Whales are our long-lost brothers."

On that rock, the one shaped like a humpback's dorsal fin, I haven't seen a harbor seal in at least ten years. In that narrow passage where the whales dove under the seine, I haven't seen a killer whale. On every rock, a seal once sat. Now perhaps the seals are coming back, but it's too late for the AT1 killer whales. The oil spill killed half of them. They are characters—some might say ghosts—in the hidden stories of every rock, every island, every bay in the Sound. New Year Island. Verdant Island. Point of Rocks. Aguliak. Herring Bay, Upper and Lower. Killer whales in Drier Bay, Johnson Bay, Elli's Cove. All along the oiled shores. The oil went high. The oil went deep. Into their stomachs. Into their lungs. Into our brains. But not deep enough, it seems, to change our minds or our ways. We thought it might, in those early days after the spill, when people all over the United States burned their Exxon credit cards. On hot days, in Sleepy Bay, oil bubbles up, as in a dream. One can't stop it. It reeks of, as Molly Lou put it, "perfume of earth's ass." It reeks of our need. For a Hummer. For a faster boat. For record profits for the Exxon Corporation. For the road to Whittier that brings more people, more skiffs bristling with fishing poles, deep and low in the Sound, to fill the empty places inside.

When I sit here to speak to it, I can't. I can't bring myself to recount it, that year 1989. It's easier to talk of aftermath. I've finally come to the place where I don't see, on the beach at Point Helen, a clean-up operation, a squadron of people in oil-smeared raingear pressure-washing the black stones with boiling water, where I don't see, on the eve of a fifty-knot storm, booms and equipment left near the tide line, even after we called on the radio to warn them it would all be washed away. I don't see, like sunspots in my eyes, those oiled booms floating around in Montague Strait for days after the storm. I don't go into Mummy Bay, where, pre-European contact, the Chugachmiut people placed the mummified remains of their honored dead in sea caves, and see in my mind a gleaming white ship (that is, if you ignore the shit-colored stain along the waterline), the one housing clean-up workers, exhausted men and women dropping spent cigarette butts into the bay. I don't re-remember the story of a clean-up boat charging around Mummy Bay, grounding on a rock, spilling a hundred gallons of diesel.

There are things one loves to write. A structure presents itself and makes the going easier, makes telling an art. Not a telling which is someone pulling the entrails out, dragging them up the beach. Your entrails, reader, your beach, our beach, the ancestors' beach, sterile now.

And this is not how I wanted to write it! It was to be poetic, sad. As in a series of present tense days, each proposing a different absence. The throb of silence in your ears, reader. Let me try again. And not to be sad, not to be sentimental. But to propose, finally, that a place persists. That there is joy seeing an eagle's nest above Whale Camp, in seeing a sea otter mom and pup scuffling at the entrance to Italian Bay. That there is joy in hearing a report of a single male killer whale and a few miles away, a group of females off Knowles Head and thinking those are probably the remaining seven AT1s. It means those two dead ones beached in Controller Bay last week probably are not two of them. There is joy in the recent past tense of seeing them all, the remaining seven in one day, knowing they survived another winter, strong, indifferent to me as they've always been, other than annoyed or occasionally granting a little hint that they know that I'm alive, sliding by the bow with a seal in their mouths, raising their heads, an eye.

The remaining mature adult male, AT6, is Egagutak. Last month, two friends and I were in the Sound. When we dropped the hydrophone over, we heard the familiar wails of a lone male transient in Knight Island Passage, glassed with binoculars and spotted him, slowly zigzagging south toward Montague Strait. I went below into the cabin to use the head, calling up to ask what Egagutak was doing. My friend Kyra said he was milling in one spot with gulls all around. "I think he's fishing," she said. It took a second before this registered in my mind. I raced back up to the flying bridge. "Let's get over there fast," I shouted. "He's probably killed a seal." Sure enough, he was diving repeatedly in the middle of a slick acre of smooth water that smelled strongly of fish oil and brine, a clean smell, tangy, metallic, almost like blood. Bits of blubber floated up, and gulls screeched and fought for the scraps, and we tried to net up samples to freeze, to prove, genetically, what he was eating. When a bigger chunk drifted up, I caught it in the net and held it between my fingertips. It was a piece of gray flesh attached to a blob of blubber. I looked at my friends. We ate that piece of seal meat.

What does this have to do with the oil spill? Nothing. And everything. Call it an act of solidarity. Call it an act of defiance. Or simply survival. A celebration. We are all here. Neil and Kyra. Egagutak and me. Here in this place. Alive and hungry for more of this.

Weathered out at the Chenega Village dock during a gale, I made a list of things that have not changed since the oil spill:

How more waterfalls form after rain, and bays are murky brown-green with a freshwater lens, and more salmon come inshore after a storm

Picking blueberries in rain, hair and cuffs soaked

Fauria crista-galli all over the muskeg meadows, green of heaven, then maroon

Dropping the hydrophone and hearing nothing

Blueberry pie

The Mad Gaffer offering ribs of subsistence, of harbor seal

Factions in the village

George line-casting for silver salmon off the Chenega Dock

Calm in Montague after a storm, water ice-like, finish matte

Humpback blows off-white strands against the blue of Montague

Deer foraging for kelp on the beach

Squalls erasing whole islands

Tall tales in boat cabins drinking tea or Crown Royal, smoking a joint around the table, stories of wildlife encounters, always hairy scary small plane trips in or out of the Sound, boat mishaps, coincidences, arguments over details between husband and wife, i.e.:

P ——: "I was on the *Monty* near Chignik when a storm came up out of nowhere, like this one but sudden, and the skipper wanted to try to get her over a bar, into the river, but there were thirty-foot breakers, and—

M——: "It was the *Rosie* not the *Monty*."

P ——: "Hey, this is MY story."

M——: "But I know for a fact it was the *Rosie*."

P ——: "You weren't even there. I was there. But if you want to tell it, if you want to make it YOUR story—"

M——: "Never mind, go ahead (but it was the *Rosie*)—"

And the truth of the oil spill is like that. You, reader, will have to ask each one of us. You will have to listen to the arguments as well as the facts. The truth is in their confluence and in the silence and under those stones and in those numbers and on those bare rocks and passages and in the storyteller's eyes.

On the radio this morning, we check in with the *Alexandra*, trade weather observations, share our plans for the day. *Happy searching*, we say. *Meet us later for blueberry picking.* We chat with our friends Roger and Marilyn at their lodge in Mummy Bay. They've known me since I was twenty-three, feeling my way around the hidey-holes and passages of the Sound with Molly Lou, following transients in the *Whale 1*. Roger says if he would have known what he knows now, he would have chained himself to the Whittier train tunnel to stop the building of the road that opened the Sound to Anchorage day boats. But no matter how easy the access, this good old Gulf of Alaska storm settled in off Middleton Island blows them all out of the Sound. One more thing I can add to my list of things that have not changed: the weather, the rain.

Later we check in with Mark Meadows on the *Ruth M* to see how commercial fishing is going. He's off today, checking his shrimp pots. It's not a good year for salmon. We check in with the salmon, trolling a bit along Point Helen, catch only the bottom when we get too close to land. Before blueberry picking, we check in with Montague Strait to see if killer whales are out there calling. We even check in with Hive, a resident killer whale. We get in cell phone range to call a colleague who's monitoring the web site where pings from Hive's satellite tag are downloaded. We plot the coordinates. Hive's up in the middle of the Sound.

Hive tells us that the killer whales' lives go on in every weather, stormy or fair, and, it seems, against the odds, for now—lucky, lucky, blessed us—they carry on. That's the present tense that matters. And yes, the oil seeps into every sentence, every observation, every breath. But the present tense is real. Reader, it's the sacred grail, the pot of gold, the pearl without price. And today, in this holy present—August 11, noon—in the lee of Knight Island, this place and these animals endure.

There is an eternal present tense. Reader, you are in it. Speak to me now. Tell me about the Sound. Does it still rain there most of the time? Do killer whales hunt along the rocks and islets, among the bergs of Icy Bay, searching for harbor seals? Do humpbacks feed off Danger Island? Have the salmon returned yet? Are they late? Are the blueberries ripe? Is it a good year? Is everyone safe? Please let me know, so I can rest and not be so afraid.

Looking for Gubbio

In dark that holds you still
In your step, so tangible
As anything, alive, unknown
You wonder what is nothing
When you cannot see a thing at all.

— Molly Lou Freeman

When I arrived in Alaska from my native western New York, in 1986, in my boyfriend's small car with a gas can strapped to its back fender, I wasn't thinking about killer whales. I wasn't thinking about fish, either, even though I'd just finished a degree in fisheries biology and was headed for a job at a salmon hatchery on an island in Prince William Sound. I was looking for wolves.

When we crossed over the border from Canada to Alaska, I imagined I was finally in a place big, remote, and wild enough. We stopped at a gas station in Tok, and when I asked the attendant if there were wolves around, he happily told me that the Department of Fish and Game had gunned down 900 from airplanes the previous winter for predator control. I walked back to the car in shock, thinking, inexplicably, of the famous Zen koan: "What is the sound of one hand clapping?"

In a dictionary of symbols, wolf is earth, evil, devouring, fierceness, familiar of death gods, devil, spoiler of flocks, stiff-necked, cruel, crafty, heretic. In non-Western cultures, like the Haida, Nootka, and Tsimshian of the Pacific Northwest, or the Sioux of the Great Plains, wolf is a symbol of power, a teacher, both hunter and hunted. When I came to Alaska, I was looking for the Wolf of Gubbio, tamed by St. Francis, himself a heretic, a holy man who claimed animals had souls. And I was looking for the wolves of *Never Cry Wolf*, the film I saw my senior year of college. I couldn't get that stark, Arctic windscape out of my imagination, or the images of caribou on the barren ground, of wolves chasing them, of their howls scraping and careering with spindrift along white, hard-packed lakes.

One of the first questions native-born Alaskans ask newcomers is, "Why did you come here?" There's a popular assumption that people who come to Alaska are running from something or searching for something abstract, like freedom, simplicity, or wilderness. When I'm asked why I came, if I'm honest, I have to say, sheepishly, that one night in Syracuse, I saw *Never Cry Wolf,* and I had to come. Of course, it's never that simple. I was running both away from and toward something.

On a winter day, shortly after seeing the film, I walked along Lake Erie's shore with a friend. I grew up on the lake in a landscape of second- and third-growth beech and maple forests, small farms, vineyards, old towns of decaying brick buildings, strip malls and billboards advertising dog races, giant water parks, and motels.

The lake was frozen, the sky a soft blue, like a worn cotton apron edged with thin clouds. Some force in the lake had thrust car-sized ice chunks up on the beach, and we wandered among them, and clambered over them, but farther out, past the jumble of gravel-encrusted bergs, the ice was flat and white and seemed to stretch north forever, to the polar cap. This must be what Alaska's like, I said to my friend, imagining the shore-fast ice of the Arctic, something I must have seen once in a magazine photograph. Now, when I return to my hometown, in summer, I scan for whale blows above the lake's chopped, cobalt surface. I forget it's a lake. The eye that searches for wolves, for spouts, for freedom, is desire's eye, and soon what it has seen becomes as necessary to the body as a lung.

In the end, looking for wolves, like looking for killer whales, is more than an act of scrutiny or listening—it's an act of patience, of devotion. It's a long story of waiting. It's a story of desire. You scan the ground for prints and signs. You search slopes for loping shapes. Wing-strokes of an owl brushed into the snow divert your attention, send you postholing into deeper snow, deeper woods. At night, you listen for howls. While you wait, you hear wolf stories from folks who've studied wolves or who've lived in places like the Brooks Range, or the foothills of the Alaska Range, or the North Slope. You hear the voice of your own longing, a trail, if you follow it, that leads your eye farther into a landscape populated as much with absence as presence.

As I sit to write about looking for wolves, as if to prove a point, two sandhill cranes, voices saw-blading, flap heavily in unison over the roof of my house, and, seconds later, a startled young moose bolts past the window, plowing through soggy spring snow, into the woods.

In 1987, Denali National Park Service biologists snared a male wolf near a dump. They strapped a radio collar to the animal's neck and released him. Using the collar's signal, they tracked his movements over the mountainous country of the eastern park. The wolf found a mate, a female about his age, and the biologists collared her, too. The pair roamed the Riley Creek and Savage River drainages, wilderness bounded on three sides by settlements and roads.

The wolf pair traveled and denned, hunted and slept in a landscape intimate and densely forested along its river bottoms, treeless and exposed to winds in the high country. Along the slopes of the Alaska Range, winter ends in June and begins again in August. Wolves share the country with Dall sheep, moose, caribou, ground squirrels, hares, grizzly bears, marmots, martens, foxes, porcupines, wolverines. Those animals that don't hibernate through winter withstand subzero cold and eighteen hours of darkness, ground blizzards, depthless snow along the rivers and creeks, and brittle, slabbed snow up high.

During a ski trip through that country with friends in March of 1992, we rested on the broad pass between Windy and Riley creeks. The wind penetrated my nylon jacket, turning my sweat cold. I pulled my hood over my head. While we passed a water bottle back and forth, our three dogs curled in hollows they'd dug into the hardpack, their noses burrowed into their belly fur. Wolf prints, elongated by wind, shadowed the snow around us.

We followed the tracks over the pass and down into the Riley Creek drainage but never saw wolves. Whenever I've visited Denali Park in winter or summer, I've looked for them. My first summer in Alaska, a biologist described to my friend and me the location of a wolf den, just outside the park boundary. For two cold, drizzly days, we huddled on sleeping pads, layered in long underwear, sweaters, wool pants, and raingear, peering through binoculars at a dark space in a hillside across the river. No wolves emerged from the den those days, but I found ropes of furry wolf scat on a scree slope during one of our walks. I lived eight years in Alaska before I saw my first wolf.

A friend of mine says our lives are dreamlike, but we don't act like they are. In dreams, we accept things bizarre, out of context, perverse. In waking life, we're unsettled by the unpredictable nature of experience. It worries us. I want to know what comes next, and when it happens or doesn't happen, I want to know why. When I see a wolf, I call it grace. When I don't, I wonder what I did wrong. The poet Rilke said, "The animal that's free has its destruction behind it and God ahead of it, and when it moves, it moves forward forever and ever, like a flowing spring." Only in dreams do we move that way. We believe that anything can happen. In waking life, we live by a different poetics, Mick Jagger howling that we don't always get what we want, get instead, what we need.

January 1996, first light, north of Fairbanks. Skiing out from the hot springs, I stop to rest on top of Tolovana Dome. A shape lopes across the trail, pauses. I hold my breath and peer at it, then realize what it is: rangy, gray, long-legged. A wolf. I stand there, long after it's trotted down the slope and is gone.

In May of 1993, the wolf pair shifted south and east, out of the park, to the country along the broad, winding Nenana River, where the highway closely follows its curves. Another wolf pack took up residence in the region the pair left behind, but they produced small, weak pups, and the dominant male was killed in an avalanche. The pack never established itself and eventually disappeared.

I've always thought of a wolf pack as permanent, but the Riley Creek pack is a shifting, ephemeral entity. There are always wolves passing through that country, but not the same wolves. They are vulnerable to starvation, to fluxes in prey abundance, to avalanches, to one another, to leg-hold traps as they move without regard to boundary markers. It's just one wilderness to them.

A few miles down from the pass, at dusk, we set up camp near the tree line. Finding its channel through the Riley Creek drainage, the wind swept down with a high-wire hum. But in the spruce grove's shelter, we felt only an occasional gust as it rustled and flapped the tent cloth. I lay on my back in the dark, listening. Like an aeolian harp, the wind described roughness, hard-frozen scree, poised snow across avalanche slopes. Poplar branches clattered softly like claws on clean ice. How does a wolf's howl begin? If I heard one, would I be afraid? Would the howl describe something foreign inside me?

The week before, I'd attended my first Buddhist meditation retreat. In the company of a dozen people, I'd spent a weekend in silence, meditating several hours each day. I'd arrived at the retreat center armed with my running shoes, diary, and one of my friends. I had a goal. I'd just finished graduate school in marine biology, and I wanted to contemplate what to do next with my life. I didn't understand what meditation was. The first evening our teacher instructed us not to write or exercise during the retreat. She'd lead yoga stretching sessions each morning, and we'd practice walking meditation on the trails and roads outside the center. When she told us not to make eye contact with anyone from then on, I glanced at my friend. Her eyes closed, she knelt in complete stillness on her bench. She was only a few feet from me, but I felt like I'd been thrown off a train in a field with no suitcase, no wallet, no map, no identity.

The next morning, our teacher instructed us in meditation, "Watch your breath. Notice when thoughts come. Watch them drift by like leaves. Just let them go. When you notice you've become lost in your thoughts, return to your breath." When she struck a brass bowl with a wooden mallet, the pure tone twanged in my skull like a plucked string, and I closed my eyes. Within minutes, my head filled with overlapping conversations, like a secretarial pool or a café—memories, worries about the planned trip to Denali Park, wondering how my friend was doing with her meditation, annoyance at someone's hacking cough, disgust at myself for being annoyed, fear that my life had no direction, knee pain, concern about my marriage, fantasies about comfortable clothes I might buy for meditation, a roiling in my stomach, a weight, like a manhole cover, on my chest. When I remembered to breathe, my head cleared, and then a large hand pushed it underwater again.

In late morning, our teacher told us to go outside and practice walking meditation, which meant concentrating on each footfall, moving at the speed of a stalking predator. I fled, stumbling up a power line trail to a ridgetop, trying not to run, my breath tearing through my throat until I finally sat down on the snow in the cold sun. The breaths turned to sobs. *How can I stand myself for the next three days?* I lay back on the snow and stared at the throb of blue sky threaded by power lines.

My head emptied out, I walked back to the retreat center, watching chickadees, listening to a slight breeze in the birch branches.

In a poem, James Wright says,

There is this cave
In the air behind my body
That nobody is going to touch.
A cloister, a silence.

During the afternoon meditations, the chattering voices broke into my head again and again, but I learned how to find the cave for a few breaths, the emptiness, what the meditation instructor called the wide sky in my mind. Outside the windows, life happened. Redpolls clung to swaying branch tips. Quick as eye-blinks, they flew from sight. I longed for their sense of purpose, the necessity of each action.

I left the retreat not knowing what to do with my life. I focused instead on the upcoming trip. Maybe, if I brought what I'd learned about meditation to my trip to Denali Park, I'd be more present and attentive, and maybe then I'd see a wolf.

My friends and I descended to the creek bottom the next day. As I looked up at the exposed rock faces on either side of the drainage, I imagined we must be in the narrow swath of the Riley Creek wolf pack's range. But mountains are staircases and not barricades to animals who move continuously. As I scanned the slopes, my friends skied on ahead of me. When I caught up, they were crouched at an open lead in Riley Creek, watching a slate-gray bird—an ouzel. It bobbed on a rock, and we waited to see if it would fly beneath the water's surface, as they're known to do. In streams with currents too strong for humans to stand in, ouzels walk with their heads submerged, hunting insects.

The ouzel stood on the rock, dipped its body up and down, stepped out into the stream, all the while singing, a string of notes rumpled like a blanket, tumbling like gravel down a bank, falling and springing up like glass beads striking ice. It wasn't mating season, and besides, its voice wasn't loud enough to announce its territory to anyone. It sang to itself as cold water slid past its stick-brittle legs. I imagined being alone in that valley. What would my voice sound like if I tried to sing? Like a croak, like a chalkboard scratch in the stillness, scaring nearby wildlife? Could my throat release some pure tone?

In a dream, I'm on a solo backpacking trip in the desert. I find a large cave in a mountainside. I know that wolves sometimes use this place. Even though I sense I should be afraid, I sleep in the cave. In the morning, I sit outside on the dusty ground. Everything is tan-colored, like deerskin. When I look up to the ledge above me, a black wolf strolls over, sits down and stares at me. I stare back into its yellow-gold eyes.

Biologists detected that the radio collar of the male wolf of the pair was losing its signal. They shot a tranquilizer dart into the wolf from a helicopter, but the dart penetrated its spine, and the wolf died. The female was joined by a black male, but he died shortly after they mated. His carcass was spotted by boaters. Seeing the collar, they assumed it was a dog's, until farther down the river, the strangeness resolved itself: the dog was a collared wolf, probably drowned.

When I decided to return to Riley Creek alone, I didn't know this story. The wolves in my imagination lived in an elusive pack, winding through spruce swales, cornering moose calves against river bluffs, gathering to chorus on ridgetops. Like biologists or trappers, I'd persist; I'd place myself directly in the pack's path. Drowned wolves, darted wolves, widowed wolves—I didn't imagine the absence of wolves.

I make a list of things I'll need to bring: matches, ski wax, sunscreen, sunglasses, food, water bottles, tent, sleeping bag, sled, extra clothes, books, diary, camera, binoculars. It's the first time I've done something like this. I've read books about women alone in the wilderness, but I've always been too afraid. It's late winter now, sunny and near freezing during the day, not too cold at night. By myself, maybe I'll have a better chance at seeing wolves. But it's not just wolves. It's clearly a challenge, to be alone the way I was at the retreat, the way I was before my marriage, to see what I am when no one's watching.

Three days later, I stand on the trail. My friend Kathy spent the first night out with me, helped me set up camp, and now I listen to her snowshoes crunching as she climbs the hill back out of the valley. The March sun warms my face. Against the white mountains, the sky's so blue, it seems scorched. Three days. Only three days. I'm already counting down the minutes.

I've borrowed a dog, Annie, to pull my sled; she'd been on the previous trip. She's a brindled Tibetan mastiff with droopy eyes and what a friend calls "lugubrious jowls." She sits next to me, holding one paw up, staring at the place where Kathy disappeared. Nearby, a red squirrel trills. I think of my Tibetan bells at home. Before meditating, I strike them together, holding them by their leather thong. I listen for the moment when the sound dies completely. When I hear nothing, except for an occasional wind gust, I flip my right ski around, then my left, head back toward my camp.

I stop at an eye-shaped opening in the white surface of Windy Creek, a good spot to fill my bottles. The snowbanks round softly, four feet down to the water. I unsnap my ski bindings, shake off my skis, and sidle to the edge on my butt. With a string tied to it, I cast a bottle into the stream. My feet knock snow clots into the water, and I watch them turn gray, then dissolve.

A crenulated ice glaze rims the opening; water flows out from under it. Passing under the rim, touching the top of the ice here and there, water makes oval shadows, like shapes of fat fish passing by. When the water emerges, its sound is glasslike, birdlike. As it ducks beneath the ice at the opening's other end, the shelf gulps it down. Beneath the ice, the water's gurgle is muted, like downstairs conversation heard through an upstairs floor. I break the ice below me with my ski pole. Shards swirl, then catch the current to the other side, where they clink against the downstream edge.

I soak my washcloth and hold it against my face. The cold distracts me from an ache beneath my sternum, the same place, during meditation, our teacher told us to return to when our minds wandered. I'm aware of each movement of my hands.

At camp, I cram bagels, pemmican, my diary, and water bottles into a knapsack because I'm determined to explore up-valley, but then I procrastinate, poking at the coals of my breakfast fire, reading a novel, *The Picture of Dorian Gray*, which, with its dank Victorian setting, is utterly incongruous with me here in the wilderness. Finally, I heft my pack and ski down the creek, scanning for wolf sign. A dog-team trail winds down the creek's middle, swerving around open leads. Since the last musher used the trail, new snow has fallen, and fresh fox tracks stipple the surface. In one spot, they veer off into deep snow, down a slight bank, then back up onto the trail.

I relax, following the creek because it's familiar. The mountains rise close and steep on either side. The way the frozen creek curves out of sight reminds me of Sterling Creek. My hometown is named for that shallow, glinting swath of water that's cut, over eons, a gorge a hundred feet deep. As teenagers, my friend and I skated the frozen creek for miles. Once, licking icicles plucked from a frozen waterfall on the gorge's steep shale side, we triggered a small rock and ice avalanche, and a wedge of stone broke the skin on my scalp. We felt like Arctic explorers. Instead of wolves, we feared reputed packs of wild dogs and bobcats and hoboes and ghosts of people who'd died falling a hundred feet off bridges. I wish my friend was with me now. I shake my head clear, shake it empty again.

At the creek's east and west branching point, a shape dashes to my left. Before I can focus, it slithers down into an opening in a snowbank. River otter, weasel, mink, wolverine? I hiss for Annie to sit and stay. She whimpers, staring at the hole. Unclasping my ski boots from their bindings, I inch up to a small cave in the bank. Suddenly signs are everywhere: a path, iced from small feet pressing down, urine stains, a scatter-plot of droppings. Crouching, I peer into the hole. I imagine the animal panting, so close that if I force another inch from my vision, I'll see a glint of black or yellow eye, breath smoke, but I don't see anything.

Except for gusts sledding down the northern avalanche chutes, the valley is still and hot in the sun. Over my left shoulder, a transparent moon rises above the ridge. Its fat, waxing edge dips toward earth, like a bowl about to spill. Annie and I eat lunch, watching to see if the creature will emerge from its hole but see only a golden eagle, circling on a thermal.

After lunch, we reach the branching place where my friends and I turned right and headed over the pass to Riley Creek. Annie and I take the east fork, so I can see someplace new. Moose tracks begin, weaving drunkenly on and off the trail, broken through crusty surface layers to weightless snow beneath. Moose-trampled willows, the color of a wolf's pelt, grizzle the hillside. Bent and bowed by snow, their chewed burgundy branches tangle together. Those branches are the tops of buried bushes, which are eight feet tall in summer. With my binoculars pressed to my eyes, I search the mountainsides. Tiny snowballs roll down the slopes, leaving slim trails.

A couple miles farther up the valley, on the edge of a cottonwood stand, I sit again and write in my diary.

> *No wolf sign. I believe that wild animals offer themselves. So what does it mean to not see wolves? I saw three eagles along the creek where it is wide. One flew out of the poplars over my head. This place belongs to the animals. Through these willows and alders, wind chimes like distant bells. Closer, it scratches through insect galls shaped like roses. Where do I fit in?*

Into a press of silence so complete I can hear static in my ears, another sound penetrates, a whine, rising, lowering in pitch and volume until I know what it is: snow machines. They're coming from another river valley, carving loops into the mountain slope, thumping on trails, coming here, into the park, where they're not allowed. I shove my journal into my pack, grab Annie by the collar, and drag her to the cottonwood stand. She whimpers, head craned back. She'd chase them, snapping and snarling, teeth bared, wolflike, but all I can think of is to hide. I crouch, heart pounding against my eardrums, as four helmeted men straddling neon green, red, and black snow machines swerve by on the dog-team trail.

For another half hour, I sit there, so their presence will be long gone from the valley, but as I ski back, I taste gas exhaust, think I still hear their mosquito-whine. Their tracks have churned up and corrugated the dog-team trail. I pray they don't see my camp.

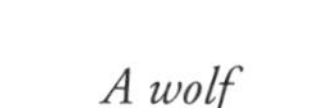

A wolf
I considered myself

But the owls are hooting
and the night
I fear.

—Sioux traveling song

Later that night, I lie in my sleeping bag with Annie curled beside me. Because it's freezing, I'm wearing my clothes. Two bottles filled with hot water warm my feet. I'm uneasy. To distract myself, I read *The Picture of Dorian Gray*, but it spooks me, too. It describes the consequences of vanity, the underside of self-reflection. More than anything, Dorian Gray fears his own demise. I think of Rilke's idea, that animals live always looking forward to God ahead of them. I live the opposite way, my possible destruction always before me. Other than the occasional pop of a campfire log, outside it's still. For all I know, the Riley Creek wolves are just over the mountain ridge, on the Sanctuary River, where sheep move on their winter range. I read a few lines, then pause to listen.

I listen for what I'm afraid of—breaking branches, a crunching footfall in the snow, grunts, the buzz of a distant snow machine. What if a cranky spring bear blundered into my tent? Or those men? Did they see my ski tracks heading into the woods to my tent site? What if wolves descended the slope near me? I saw a wolf-killed moose once. Its body lay twisted and disemboweled in old snow trampled by wolf tracks. Blood stained the willow branches, broken by the moose's staggering. Urine yellowed the perimeter. There wasn't any pity in the scene. From somewhere, perhaps a novel, or a family story of life in Latvia, an image arises, of a horse-drawn sleigh careening through dark woods, hounded by wolves. I lie back in my sleeping bag, my arms stiff. Some Native people say that the *yeda*, the wolf's spirit, is so powerful, so sacred, its name should not be spoken.

In *Of Wolves and Men*, a book I read soon after I arrived in Alaska, Barry Lopez describes a tribal initiation practiced by the Nootka, Kwakiutl, and Quillayate people of coastal British Columbia in which young people are symbolically stolen by wolves, undergo terror, and "emerge wolflike." The initiates must "be as fierce, as brave, and as determined as the one who is the greatest hunter of the woods." I've seen wolves carved into Tlingit totem poles in southeastern Alaska. Poles conveying a person's familial lineage incorporate humans and animals whose facial expressions share a similar, fierce, supernatural power. On one pole, near the bottom, a human hung upside-down, half-swallowed in a bear's mouth.

I huddle next to Annie, and I'm afraid to hear wolves. This is what I am when no one's watching. A mouse, not a wolf. Prey, not hunter. I feel suddenly the impulse humans have to build sturdy walls against the wilderness.

The next morning, as I take apart my tent and drape it across spruce trees to dry in the sun, my fear seems absurd. Inside my head, a voice ruminates. Is it instinctual for humans to feel like prey at night? Is it because I'm a woman? Is it because I've always been afraid of the dark? Will I encounter wolves when I'm no longer a stranger in wild places, but feel native, like I belong, when I no longer fear death? Is this some kind of failed vision quest?

As I walk back to my campfire, I spot fresh animal tracks in the snow, the paw prints broad and big, the snow behind them swished by a tail—a wolverine, I think, with excitement. Annie snuffles her nose into a track, jerks up her head, sniffs the air. I grab her collar and whisper, *Heel.* Holding her close, I step through deep snow, following the meandering tracks to a stand of cottonwood trees. Around each trunk, shadows pool in blue moats. Suddenly, an animal's shape coalesces out of a dark lump at a tree base. It's curled tightly into a ball. Annie lunges forward, and I yank her back, whisper, *No*. Slowly, the animal untucks its head. Black, flinty eyes glare at us. Then it buries its face into its chest again, clamping its paws over its forehead. I laugh out loud. It's a porcupine.

The night before my marriage ceremony, a porcupine wandered through the wedding site, an uninvited guest, even then. I want to be wolflike, but here's what's given—my totem—this creature balled up in self-defense. A rustle startles me, and I look up. Three gray jays fall through the air, twirling. The sky is a blue shell, and sun glints on the feathers of a golden eagle riding currents four hundred feet high.

Perhaps life is dreamlike, not an unfolding, in linear progression, from desire to fulfillment, but a series of juxtaposed scenes, like these: tent, porcupine, gray jay, golden eagle, sun. In absence of something—the wolf, say—there's no disappointment, only surprise.

I suddenly recall that my friends Ginny and Celia, both in their eighties now, taught me this same thing. I was standing at their door after a visit when Celia gesticulated with her hand. In her excitement, she couldn't speak. Finally, she blurted: "Look, *look*!" Her eyes fixed out the window, she grabbed my arm and pulled me toward her. Twenty feet from us, a young bull moose stood at the bird feeder. Sprouting an elegant rack in miniature, he bent his head sideways, shoved his enormous snout, soft as a chamois bag filled with pudding, between the feeder's roof and its base, his fat tongue licking at the sunflower seeds. Ginny grabbed her camera, and both women chattered and pointed.

"My, will you look at that." Laughter. "Oh, look at you. Who are you? We haven't seen him before. Do you think he's the young one we saw last fall, with the antler nubs? Do their antlers grow that fast? Shhh. Don't move."

There it was. The two women had lost consciousness of themselves. They'd lived in Alaska fifty years, moose an almost everyday part of their lives. Ginny was charged by one on a ski trail near their house. She'd dodged behind a tree and fended it off with ski poles. They'd seen moose hundreds, maybe thousands, of times. They'd constructed a ten-foot-high fence around their garden to keep them from devouring their cabbages. They'd built their own cabins, guided rafts down wild rivers. They'd flown airplanes all over Alaska. They'd seen wolves, musk oxen, enormous caribou herds. They'd been to Glacier Bay, the Arctic Ocean, the Brooks Range, and the slopes of Denali. You might say that in fifty years, they'd seen it all, but apparently not.

A camera flash reflected back at me from the window glass. Celia crouched, giggling. The moose drew its head out of the feeder. His ears, big as aluminum flour scoops, rotated in all directions. I saw a tiny frost flower at each eyelash tip.

When I return home from my ski trip, I buy myself a silver bracelet engraved with abstract Haida wolf symbols. I wear the bracelet every day. A few months later, when I pull off my jacket, my wrist feels strange. I look down. The wolf bracelet's gone.

Years later, I hear the rest of the Riley Creek wolf story. In October 1994, a pair of wolves appeared in the area of Jenny Creek, where the previous pack had come and gone. They took up habitation, raised three pups, roamed the Sanctuary and Riley creek drainages.

The lone female wolf bore a litter of pups fathered by the drowned black wolf. Soon after, one of the Jenny Creek pups joined her and the cubs. In early May, they were all found dead in snares left out for a few days after the close of wolf trapping season, outside the eastern park boundary.

In nature, nothing's a closed circle. In a scene in the Macedonian film *Before the Rain*, graffiti is scrawled on a city wall: *The circle is not round*. New wolves will penetrate the Riley Creek drainages. They'll set up dens, raise pups. They'll bring down moose, or they'll die of starvation. They'll place their feet in leg-hold traps, their necks in snares. Someone, somewhere, will adorn his parka hood with their fur. I imagine my bracelet, hanging from a branch, or buried under a pile of bear scat.

This is my wolf story. I never found the wolf of Gubbio. I never found the familiar of death gods, either. I've never heard a wolf howl. I've never seen the barren grounds. Since 1986, I've seen wolves three times, never when I've been searching. It's always been a lone wolf, and it's always been silent. Nonetheless, I've read about them, looked for them, protested against wolf control. I've studied their oceanic counterparts. A Haida legend says that once, some wolves began swimming out to sea and killing whales. Every day they returned to the village with a dead whale, and after awhile, the beach became crowded and foul with rotting carcasses. To punish the wolves' greed, their "God" whipped up a huge storm, so the wolves couldn't come back. They had to stay out there.

It's Sunday morning, early December. I drive ten miles through a blizzard to meditate with some friends. Creeping along in four-wheel drive, wipers slapping, I listen to the Alaska Public Radio news, which reports that wolves are so plentiful in the Alaskan interior that they're competing with rural hunters for moose. The Department of Fish and Game is getting ready to send out the helicopters and the gunners once more. "Not again," I cry out loud. "I can't believe we're going to have to fight this all over again."

An hour later, I sit in my friend's meditation room. On his low shelves perch framed photographs of gurus and teachers, and a snow-dusted Buddha statue squats outside the door, looking in at us. Spruce-studded hills roll to Cook Inlet and the Gulf of Alaska.

Wind leans into the house. Snow billows across the meadow, flakes falling thick and fast. The ocean is streaked, gray, ripped-looking. Toby starts a tape so we can chant in the Indian Kirtin style. Men chant, then women, and we reverse roles, then sing both parts. I'm nervous about singing out loud, but my friends' strong voices give me a blanket I can hide under. When I close my eyes, I feel sound begin in the center of my chest, where it's tight, like a wrung washcloth, and I push the sound out slowly. My voice grows louder until I can hear myself. Wind gusts shriek around the house, the world is wild, and I think of wolves, who survive out there, without triple-glass panes, without timber frames. I sing in Pali, a language I don't know. I imagine my voice sneaking under the door, weaving into gusts, finding that place, that wilderness, where the wolf howl begins.

Six Hundred and Fifty Pieces of Glass

And we
who always think
of happiness rising
would feel the emotion
that almost startles us
when a happy thing falls.

— Rainer Maria Rilke, "Tenth Elegy"

From the window of my sister's home on the California coast, I could see all the way to Point Reyes, eight miles away. The foggiest weather in the world was there. The lighthouse opened and closed its eye. Every morning, out to sea, a fogbank smudged the horizon, suggesting a distant mainland shore, yet I knew, if I set out in a boat from there, I could follow the planet's curve halfway around the world.

I went there in December for light. Where I lived in winter, in interior Alaska, the sun drooped low over the horizon, like a yolk, for four hours a day. The air was blue. I learned the corresponding gestures over fourteen winters spent there, my body's gestures, my mind's. The gestures of trees, animals, clouds. Sometimes I turned my back on the darkness and fled south. Sometimes that was the only gesture I knew.

⁂

In winter I heard voices, needle clicks, knitting inner to outer, as my mother knit me two-layered mitts for winter mountaineering trips:

A neighbor's voice. *Did you ever consider lithium? My husband's been on it for ten years.*

A husband's voice. *I think you're manic depressive. You're so up and down.*

A sister's voice. *I get into bad moods, too. It's just that yours are so extreme. You get in there and stay in there.*

"How long have I been this way?" I asked her. She laughed, and I started laughing, too.

"As long as I've known you."

A cousin's voice. *You have to be careful. Depression runs in our family. Three of our cousins in Latvia committed suicide.*

A friend's voice. *It could just be something chemical. It doesn't have to be this hard.*

A lover's voice. *Somehow I just don't believe it's chemical.*

A therapist's voice. *One day you'll walk in here with wings.*

A father's voice. *Palietz klusa.*

Become quiet. A gesture, that is, with the subarctic winter, perfectly in tune.

In winter in Alaska, darkness bows down to the earth, holding her red skirt by the tips, a black-toed shoe pointing out.

On the California coast, fogbanks bowed to wind, to the day's heating up. Wave flecks dyed the sea indigo. Fog dissolved to smoke-like wisps. Edges presented themselves everywhere: between green and blue, between wave troughs and their breakages—stark as icebergs—between my body and its gestures. I sat and watched events—clusters of birds, trawlers—play themselves out along the horizon.

The more I watched, the more the sea suggested aloneness, which is something one avoids, I suppose, at all costs. As a wave confronts the shore, always retreating.

I stared at the ocean, thoughts pouring down a slough into its roiling, before closing my eyes to meditate. I breathed in. I took a sip of seawater into my mouth and swallowed it.

Since childhood, I'd felt a sin, a poison inside me, something that held me apart from others. If I built a pure space around me—a kind of perfection—would I be pure?

Daily list, age 36:

write
run
clean cabin
play oboe
do yoga
walk dog
meditate
take vitamins
eat right

Daily list, age 15:

exercises
practice oboe
homework
Latvian homework
piano
make reeds
write in diary
pray
don't gossip
stop brooding
attain ideal weight

Off Bodega Head, every few seconds, the foghorn repeated. It sent out a call for which there was no response. The absence of response suggested the sea's extent. It had no name, no gesture. It echoed the uninflected gesture of a wave's lap on shore.

When I looked out past the window glass, the view of the sea dispersed me like a wave's crest in wind.

I dreamed of a woman swimming face up and head first through a tunnel. She held a packet of letters in her hand. She was trying to reach a destination, but the journey was long. Her body began to dissolve. Her skin streamed in tatters. Baby sharks followed her, and she couldn't get away. Halfway through the journey, she was dead. She floated, ribbons of flesh and intestines clinging to her bones.

⁂

In interior Alaska the landscape was snow-covered for seven months a year. Like an ocean, winds defined its surface. Gusts carved gullies and stiff sastrugi waves. When snow fell without wind, the landscape undulated, under swells. Its surface gathered and emitted light.

During the coldest weather, fifty-below, the sky was so blue and taut I wanted it to shatter, releasing me.

⁂

I walked along the beach during the day while my sister worked. One day, three men in hip waders cast lines into the surf. What kind of fish, I wondered, would congregate in that roil, going nowhere? I wanted to see a fish come out, to know that something existed beneath the foam and undertow and danger, but I kept going. Half a mile farther down the beach, I saw two dark, large-bodied scoters, wave after wave striking over their heads. They ducked down, popped back up, paddling faster, until the next wave came.

⁂

The woman in the dream was my friend C——. I get to this point and stop writing. I make a circle around my cabin. I trim my nails. Open the refrigerator, scan the shelves. I sit back down and confront the page. Approaching her the way a sanderling approaches waves, keeping just out of reach, keeping my legs from getting wet.

⁂

I'm not quick to make friends. I was intimidated by C—— at first. In grad school, flannel, jeans, and t-shirts were the norm. In her red beret, long jumpers, and black lace-up boots, she was a splash of color against the cinderblock walls. She was five-nine and statuesque, dark brown hair and eyes, thick black eyebrows. Unlike me, she carried her height, her body, with assurance. She studied limpets, and in free time wrote poetry, sewed, skied. Someone, you might say—and people did—who *had it all.*

~

When a wave retreats, it leaves a seam, foam speckles, holes in the sand, a blue, luminous stain. Sanderlings skitter along the aftersheen, prodding the sand.

"I want to see the lighthouse," I told my sister. We were running on the beach with her dogs. Like the sanderlings, we followed the waves' sinuous edges. Bulkier willets strode over their reflections, pausing to poke, to overturn shells. Lean and leggy like our shadows, the sanderlings ran toward retreating waves, raced up the beach, returned.

She told me to wait until my lover arrived, in a few days.

I told her I'd been writing Christmas cards while she'd been at work. Having skipped last year, I had to tell everyone about my divorce. It felt like going to confession. Shame was involved in it.

"Then why do you go there? Why do you bring yourself down?" she asked. "Your life's moved on. You should be happy."

A thin finger of land pointed far out to sea in the haze. At its tip, the Point Reyes light flared and flared.

~

She showed up in my morning yoga class. In the locker room afterward, she walked toward me, naked and friendly. After a few classes, we began to talk. Odd things revealed themselves, things I remember fifteen years later. A plum tree she painted for her father that winter. Her favorite songs: "Roseville Fair," "Gone, Gonna Rise Again," "Satisfied Mind," "The World is Always Turning Toward the Morning." She told me that she was afraid to sing out loud, ashamed of her voice. "I'm tone deaf," she said. "I mean I *really* am." I told her, "We should sing together, get over our fear."

One morning, as we dressed, I was hurrying. I had an appointment.

"What kind of appointment? "

I paused. "It's a therapy appointment. For depression."

She told me they'd put her on lithium after she'd tried to kill herself five years before. She'd stopped taking it when she came to Alaska. Her mother was schizophrenic, lived in a halfway house, had lived on the streets when she was a girl. Once, her sister had tried to kill herself, then changed her mind. That, she said—the pain she felt—decided it. She'd never take her own life.

I was reading a story about a woman who lived alone on an island, tending a lighthouse. She had no name. In the end, disappearance was her gesture. The lighthouse beam illuminated her once. On the next pass, she was gone.

The foghorn sounded. It tapered to nothing, like a flame. One of my oboe teachers taught me to *create* sound that way, from nothing, so a listener wouldn't know when it began. I thought of the boreal owl I heard outside my cabin in February:

bo-bo-bo-bo-bo-bo-bo   bo-bo-bo-bo-bo-bo-bo

bo-bo-bo-bo-bo-bo-bo

I listened closely for a response. Silence. Now the calls came from farther away, extending the range of desire, for confirmation that *I am*. In the night, do owls perceive horizon? Do they need to? Or is it only we who fix our frame of reference by horizons, by obstacles, navigate not by echoes or emptiness but by stars?

January 1990, just before her twenty-eighth birthday, C—— told me she'd tried to kill herself over the weekend. She asked me not to tell anyone. She'd swallowed a bottle of lithium, then called an ambulance. After a night at the hospital, she was released. I kept my promise to her.

I didn't understand isolation's extent. I didn't understand how, encountering vastness, points of reference could be fixed so close, they'd become mirrored walls, tiny rooms. How the mind, in isolation, reflects back upon itself in jagged pieces. How strong the persona is, how tall, how upright she might appear, while inside the mind a cage of blackbirds rattles, while inside the mind, bells clang, voices whisper, the birds are trapped, while inside the mind, the windows have been painted black. How the structure persists in the face of this.

Disappearance is also a gesture.

The sea freezes and thaws and shovels ice pans against the shore in its ebbs and flows. It drowns sailors. Sea and weather exist as fluxes, reminders of what endures, and of what does not.

A wind blew up the coast some days, driving the sea toward the Point Reyes light. There, the surface static glittered unbearably under the sun as Point Reyes faded into a pale blue shadow. The sea appeared to shift, obliquely.

As I was observing the shift, noting it, my lover called. I told him, "I don't think we could stand it if we really faced our solitude. I think that's why there are stars. I think that's why we perceive a horizon. Otherwise, we'd spin right off the planet."

"You mean in a poetic sense?" he asked.

"No, I mean literally. We'd all go insane."

Of vastness, the philosopher Gaston Bachelard says, "Here we discover that immensity in the intimate domain is intensity, an intensity of being . . ." The experience of the immortal exists within those moments, in spite of the reality so evident, that they, too, are ephemeral. The gesture of effacement swipes them away.

February 8. The night she took her life, the full moon hung metallic in the forty-below air. She bought a knife and drove herself north, up the Elliot Highway, stopping at the Hilltop Diner for coffee. She drove back toward town, but turned right onto Old Murphy Dome Road. She parked her truck along the oil pipeline corridor and walked, first on the hard-packed trail, then veered off into thigh-deep snow.

Old Murphy Dome Road, a single track, washboarded in winter, cut an arch through alders. The branches scraped the windows as I drove. It was the day the helicopter lifted her body out. In the end, after slitting her wrists, she'd fallen down a cliff. The police told me her truck was still parked along the road, but it wasn't. I drove what seemed too far, turned around and stopped at the pipeline crossing, my car idling, wondering what to do. A pick-up pulling a log behind it rattled to a stop next to me.

"Is everything okay?" he asked me.

I told him what I was looking for.

"I'm the one who found that truck," he said. "I knew right away there was something wrong. I recognized it from the description in the paper. I saw her footprints leading away."

"Did you follow them?" I asked.

"No," he said, staring up the road. He expelled air, what could have been a laugh, but there was no smile in it. "Never so brave."

I didn't tell you the rest of the dream. There was more. I went to the place in the tunnel where she had died. I went there to finish the journey. Like her, I swam face up, head first. I held the letters in my hand. The sharks came and tore flesh from my body. The fluid I swam in was corrosive, like tears or snot. I could feel my skin dissolving away. It slowly grew lighter. I climbed out of the tunnel, into a huge building devoted to science. I went into the bathroom to clean off the slime. I put on C——'s clean, white, translucent blouse.

My lover and I drove to the lighthouse. We could see it miles distant, a white tower rising above a long green field stippled white with swans.

I held his hand hard, as though I could arrest the falling I felt inside me, even though I'd practiced meditation and studied Buddhism, the religion of letting go. Had read carefully, but not happily, the book called *When Things Fall Apart*.

Every morning in California we sat in silence, side by side. Afterwards, I put my hand lightly on his back. I watched his chest rise and fall, and I breathed with him. Something loosened inside me and I floated. Was that letting go? When he watched me breathe, I felt my skin harden and gnarl into bark. I felt branched, rooted, separate. When I opened my eyes, I said, "I'm a tree."

For many years, every February after she died, something happened. Light illuminated her figure, and then it went dark as I dove underwater. I surfaced to teach my classes and sit in front of my students, a plastic doll. Inside my head, birds banged around. When a class was over, I rushed to the bathroom, shut myself in a stall and sobbed. I lived inside the plastic doll body. No one noticed.

As long as I was moving, I outran it. At home, I sat on the floor and stared at the door. A voice said just go outside, just do it, just walk out into the snow. But I was rooted. The voice pinned me to the floor. Never so brave.

I walked out to the beach and pressed my back against a dune, took off my shoes, dug my toes into sand, let the cold sea foam numb my skin. Gray wind tried to lift me up. How could I not hold on?

I imagined my lover and me on a tightrope. We each held a ten-foot-long pole. The voice over my shoulder caught me off guard. It whispered of failures, of sins. Its disembodied head floated. I leaned and fell.

I landed on another tightrope. My lover was there. I stepped forward, balancing with my pole. I lived in the center of that step. I looked down. There was nothing there. I leaned too far and tried to grasp his hand. I fell, pulling us both down.

I landed on another tightrope.

We climbed the iron staircase round and round to the fresnel lens. There was no lightkeeper, just a docent who told us that the lens was made of 650 pieces of glass, weighed two tons. It floated on a mercury bath. Before electricity, the lighthouse keeper cranked a weight every four hours to keep the clockwork mechanism rotating.

Through glass arranged in slats, like circular Venetian blinds, I peered at the light mechanism. Once, an oil lamp sat on the pedestal, emitting its smoldering yellow flame. The configuration of glass focused its weak light into a beam to reach a mariner's eye eighteen miles out to sea.

I photographed the glass from many angles. Alone, each thick piece was nearly clear. When viewed obliquely, in series, the pieces were Coke-bottle green. I photographed my lover through hundreds of pieces of glass. I thought, *I've been here before.*

I crawled under the lens. Above me, the ceiling was cracked. The lights tilted on their pedestal, bolted to a metal stand, their filaments broken, the wires cut, the pedestal messy with pieces of plaster and dust. Upon its mercury bath, the fresnel lens sat immobile.

I opened the door to the balcony and leaned outside to where wind buffeted my head. There, a new mechanized light revolved. One hundred feet below, swells rose into shallows, curled and broke, curled and broke, advancing always.

As we came away from the white tower, I stared up. "Coming to this lighthouse is the best thing that's happened to me in a long time," I said to my lover, but I didn't know why.

Prayer is a habit of language and of thought. As a child, I was afraid of the dark. I was afraid of being possessed by the devil. Prayer was my antidote. I prayed to not be depressed, to not be moody. I prayed for happiness. I prayed for perfection. I believed if I was pure enough, I would know God.

Once, to get me to sleep, my brother said, "Hold your hand out and close your eyes. If you concentrate real hard, you'll feel God take your hand." The gesture became a habit, and I still sleep with my hand like that, cupped full of nothing.

Journal, March 3, 1980, age sixteen:

> I have to figure out these troubling emotions I'm having if I am to know myself. If I can be sure in the knowledge of myself, then I think my life will become a lot easier and happier. The first, and most fundamental of my problems is loneliness. I have this feeling many times, and now even when I'm alone I feel lonely.

Driving on a deserted highway in Alaska with a friend, in winter, at night, she told me about visitations of the Virgin Mary, that she'd appeared to twelve Croatian children.

I looked at the rearview mirror. "Look," I said. "What is that?" Light dappled and danced on the back windshield, like television static. It streamed across. Then it went black.

I buy Virgin of Guadalupe candles. They burn all night.

My lover lit a candle. He placed it on the floor between us. I closed my eyes. I heard the voices ebb and flow. I pointed my foot and placed it carefully on the wire. I'd read that a single moment of consciousness is a miracle. Briefly the boundary dissolved. As the bark peeled away, the rooted tree moved its branches. Poet James Wright wrote, "The branch will not break." But the branch did break. I floated in emptiness and knew that my lover—that everyone—floats there too. There is no bottom. But we don't fall.

At the lighthouse museum, I stared at a photograph of a tower in France in a storm, a breaking wave engulfing it. The keeper leaned outside the door, camera poised, tiny to the wave.

I wanted to possess that image. I wanted to buy it, to take it home. I wanted to be the man in the tower. I wanted to witness the storm from within the tower, to wear the storm like a cape around my shoulders, to open the door and step outside, to cast my net, send my signal toward what does not reply.

I showed my lover photographs of the fresnel lens. In one, his image was flattened and broken in pieces behind the concentric circles of glass.

"It's the lighthouse keeper's ghost," I told him.

One-dimensional in the fresnel lens, he watched me, through 650 pieces of glass.

~

C—— disappeared. The night she took her life, we'd made a plan for her to come to my cabin to spend the night. Her parents had flown to Fairbanks to take her back home with them. She'd been released from the hospital that afternoon, after her third suicide attempt. When I got to my cabin door that evening, I saw a note pinned to it. "If you know where C—— is, please call her father."

I didn't have a phone, so I waited at my neighbor's studio. To distract myself, I played the piano. Several times, I thought I saw headlight beams and ran outside. I ran to the driveway's end, but there were only shadows cast by the moon. At two in the morning, I lay down on the floor and closed my eyes. In my mind, I saw her haggard face. She said, "I'm at peace now." I fell asleep.

But the next day I searched for her, driving up and down the Elliot Highway, once even driving up Old Murphy Dome Road, but not far enough.

A day later, a knock on my cabin door woke me. My neighbor's voice called out, "C——'s parents are on their way over." I pulled on my clothes. I paced the cabin. Something that couldn't be stopped, a wind, rushed at me from far away, a williwaw bearing down a mountainside. When her parents walked through the door, the wind knocked me down. I fell to my knees.

~

I want to see and understand the phenomenon of depression, the thing itself, as poet Wallace Stevens says, the reality, the organic center, free of metaphor.

But it's only by metaphor that I understand. In my lighthouse, the windows are blackened. I'm trapped inside the glancing light and beating wings.

A scientist once told me about a question she'd been pondering. What would happen if there was a perfect hollow sphere whose inside surface was a mirror. If it had a tiny door, and one shined a bright light inside the sphere for an instant, then slammed the door so that it was completely sealed, would the light left inside ever die? Would it keep reflecting itself over and over, forever?

~

Once, an oil lamp was there. The wick needed constant trimming, the chimney constant cleaning. Soot built up on the fresnel lens, and it took the lighthouse keeper eight hours to clean the 650 pieces of glass. Lives depended upon that maintenance—a kind of perfection, a species of devotion. It's my gesture now.

I must work incessantly to clean the 650 pieces of glass, to rub the glass clean, so the beam can escape. I open the windows so the birds can come and go at will, so they're not trapped inside or restless. I sit on a cushion and close my eyes. The lighthouse beam sweeps the horizon. I think of the woman in the dream, the one who disappeared. How the beam searches for her, always.

❧

At night, in my ear, the foghorn repeated with the rhythm of wave beats, white curl of the sea edge, pleats in the gray of the sky.

❧

In winter in Alaska, darkness bows down to the earth, holding her red skirt by the tips, a black-toed shoe pointing out. She is showing me what to do. So I might one day respond to the gesture of darkness by bowing.

Ghosts of the Island

Listen, you say, to the waves on the beach.
Having been here day by day,
the water almost disappears by constancy.
You step back from the fire and I follow
always. When it releases us
it is almost as though
we aren't here at all.

— Molly Lou Freeman

I guide my boat north toward the island. The landmarks—coves, mountains, channels—are as familiar to me as my own hands, but still I'm careful. Off Lucky Bay, I scan the water's surface for the rock that once, ten years ago, I mistook for a whale's back, sounding. The humpback rock is the farthest out of ledges at the southern tip of the island that break the water as the tide falls and submerge again as it rises. At the highest tides, only the farthest rock shows, and water laps around its fin but never covers it completely. At lower tides, the rocks slit the water's skin the way tips of mountains pierce a fogbank. Cormorants and seals plaster themselves to the exposed surfaces.

I haven't been back to the island since last fall, when Molly Lou and I closed down our tent camp for the winter, left it behind for good. Everyone around this part of Prince William Sound knows the spot as "Whale Camp." At night, by the light of a gas lantern in the tent, I took a pen to paper and mapped the route our small boat made down the passages in search of whales.

I won't be living there anymore, and Molly Lou's not with me. I'm using the *Whale 2* now, a converted gillnetter large enough to live aboard, much more efficient for the research than daily runs back and forth to the island.

For now, the island's southern end conceals the camp from view. But it's near, and something, a wind wrinkling the water—memory inside me—triggers a lonely feeling I push back down.

I make for the humpback rock, knowing I can round it closely. Navigating by instinct, I fix my position with the south end of the Pleiades Islands and stay offshore until I reach the point. Another rock between the point and the entrance to the channel behind the island emerges at only the lowest tides. I've seldom seen it. The *Rejoice*, a salmon boat, ran aground on that rock a few years ago.

Since the rocks drop off a hundred feet or more here, I pass closely enough to count the archaic shapes of cormorants clustered on them, their javelin wings outstretched to dry, their long necks craned to follow the point of a wingtip, as if their bodies have been pressed between panes of glass. I'm so close to the rock and to the birds that I can see the orange patches on their faces, the deeper blackness of their glinting eyes, the iridescent green—color of a fly's body—on their feathers.

As I leave the point behind, the wake of the boat slaps sharply against the rocks. I look back, startled, and then a weird image enters my mind. I imagine bringing the boat around the last headland, expecting to see the shoreline empty, blank. Instead, the tents perch birdlike on their buckled platforms, their white canvas flapping in the wind. From the alders behind the tents, Molly Lou emerges, carrying buckets of water drawn from the pond. I recognize her small form, moving purposefully and fast. The water sloshes, its weight pulling her arms straight down, propelling her forward, her knee-high rubber boots digging into the beach stones, a look of concentration in her brown eyes. And I'm there, too, crouched in front of a beached log, tending a fire. The boat comes closer, but they don't look up. They don't see me at all.

I shake myself out of the image, but like a dream's aftermath, the disconcerting feeling remains, and it stings in my armpits. What if Molly Lou and I still lived on the island, held within some parallel reality I can't enter, the strongest part of us left there and the part of me here on the boat, a simulacrum, a stranger? Which one of us is real? The woman tending the fire or the one seeing the beach from a distance?

Then the actual beach stretches into view as I round the last headland: rock outcrop, pebble beach, rock outcrop. Above the tide line, ramshackle tent platforms tilt toward the water, ryegrass obscuring the plank walkways, the rounds of wood jammed beneath the plywood floors. There are no canvas tents, no figures of Molly Lou or me. And later, when I walk up the beach, there aren't any footprints, except for some deer tracks pressed into the black sand exposed by the tide.

I step onto the walkway. Fishing web we tacked to the planks adds purchase to the rain-slick wood. The tent platform has decayed more over the winter, and it creaks and bends as I walk on it, even though we nailed thick pieces of plywood over the rotten wood in the corner last summer to patch the floor. Ducking beneath the hip-high railing that edges the platform on three sides, I see the cache tucked behind alders in a hollow in a rock ledge. Someone has been through it over the winter, the tarps dragged away, coolers and buckets opened and filled with rain. Rust-streaked pots and utensils waver under a foot of water. I remember the hours we spent last fall, meticulously storing everything away.

Between the two tent platforms, the intruders have left their own trash—Spaghettios cans, sardine tins, spent rifle cartridges—and I pick them from the fire circle and pile them on the platform. They leave a slime of wet ash with an acrid smell on my fingers.

A squall moves in, dropping mist on Knight Island Passage. In our own pit, erased now by winter wave action, Molly Lou and I cooked flounder or rockfish we'd caught. Our last night on the island together, we built an enormous bonfire to burn our summer debris, wax boxes that sizzled and billowed, releasing a spiral of sparks into the dark. It seemed we burned all but the essential between us and the island, all but the white ash of friendship in that fire.

Crouched at the cache listening to the slap of waves against the shore, I hear a line Molly Lou wrote about that time: *I keep this place where you have been, coaxing the fluxes.*

The island takes the beach back for itself. What, in the end, remains of our time here? The Sound is strewn with such ruins: fish camps, abandoned mines and canneries, settlements. Where are their stories? A herring saltery once operated in Port Audrey, at the head of a deep bay near here. One day Molly Lou and I searched there for the cabins marked clearly on our chart. We found only piles of fireplace bricks.

Across the passage from our camp, at the site of the old Chugachmiut Native village on Chenega Island, only the old schoolhouse still stands, built on a hillside overgrown with salmonberry bushes—plants of disturbed places, "waste places," the books say. In summer, thimble-sized berries dot the branches, some the color of sunshine, others of blood blisters.

The white paint of the school has chipped away, the wood beneath grayed, the windows broken so nothing reflects. From the boat, my eyes travel into those openings, trying to imagine the clear fir planks upon which children's feet fidgeted. Long ago, looters hauled the floor away.

The schoolhouse's elevation spared it from a series of massive tsunami waves that, during the earthquake of 1964, sucked all of the water out of the bay in front of the village before slamming it back, flattening the houses and wharf, killing 26 of the village's 120 inhabitants. The wave flung boats into the forests, shaved stands of trees to a field of broken masts. It's thought it came from an enormous chunk of ice that broke off a glacier face and plunged into the fjord during the quake. Chenega Island rose five feet. It shifted fifty-two feet to the south.

The survivors left the village, leaving the dead behind, a plot of crooked Russian Orthodox crosses marking their graves. A Chugachmiut woman told me that owls living among the caved-in houses are spirits.

Leafing through a book of Chugachmiut legends, I study the black-and-white photos of the village in the 1940s. In some of them, I recognize the child-faces of friends who now live in the new village, several miles from Old Chenega. One photo, in particular, haunts me. In it, men cluster around a rough wooden table, all dressed up for a formal occasion in black jackets and white shirts, the wide collars tucked into their lapels.

On the table, the flame of an oil lamp flares, sun-like, and nearby, a candle shines a similar sun, as if it were the reflection of the lamplight, or something about to happen. A calendar, the dates blurred, hangs on the wall. Against the door, a disconcerting shape slouches, perhaps just a jacket hung on a nail, but resembling a figure.

The stovepipe rises behind the men's heads. Some rest their hands on the checked oilcloth, some eat bread, and one drinks tea from a cup so white it looks like a cut-out. In the foreground, three men lean forward slightly on their chairs, their foreheads creased from the eyebrows arching up. In each eye, a bead of light glints. I examine these glints, to see if there is anything reflected back, for the miniaturized image of the person with the camera, but there's nothing, just a white blur, like the lamp's flame.

In those eyes, I search for the wave, already coming at them from twenty years away. In their faces I imagine some knowledge of a time when they would be but ghosts of the island. And yet the real moment depicted by the photograph is plain, unremarkable. Men and women resembling them live in New Chenega, the mailboxes and the crosses printed with the same names: Evanoff, Kompkoff, Eleshansky, Totemoff, Selanoff. One of the faces belongs to my friend Mike. I recognize the younger version of him mainly by his ears.

Now an elder, Mike survived the tsunami by escaping to high ground above the village, where he and others huddled together through the night, nearly every family having lost someone. Only in recent years has Mike wanted to visit Old Chenega. Some survivors won't ever go back. The ones that do won't go there without a priest. I try to imagine a feeling immense enough to turn them away from that place. I think of the island, the one across the passage, where I lived.

Big tides or heavy rains carry glacial ice out of the fjord. That ice, gliding past the schoolhouse and sometimes aground on the village beach, is also a kind of memory.

I have dreams about leaving the island. In one, I stand on the beach facing the passage, watching the Pleiades Islands float and blink. Time flattens, and the tide neither rises nor falls, all day water coiling over the same stones. I lay out a tarp to dry in the sun, then bundle all the clutter of my life onto it, fold the corners toward the middle, wrap it with rope and then roll it to the edge of the water, to a waiting boat, up a gangplank. I shoulder my knapsack and climb aboard as the boat pulls anchor and chugs away. Standing on the back deck, I watch the beach slim with distance until everything specific about it is gone. I force myself to turn away, to avoid bad luck, but I can't stand it, and when I look, I see myself—a stronger, whole self—back there on the island.

One July, Molly Lou and I returned to camp after only a few hours on the water. The day-breeze had increased to an unusual twenty-five knots, tearing off wave tops, making the water surface look like broken glass. Our faces itched with salt. We were happy to have an excuse to spend a sunny day on the beach.

For an hour, we sat on a log in front of the tents, our bare feet dug beneath the hot beach stones to the cold wet sand underneath, talking and tossing rocks. We invented names for parts of ourselves: the critical ones, the clumsy ones, the mindlessly dutiful ones, the brave ones. At noon, we ate tuna fish and cheese wrapped in what was left of old tortillas.

Neither one of us could sit still for long. We're both frantically industrious by nature. On stormy days in the wall tent, we wrote stacks of letters, transcribed data, labeled film, read poetry, baked bread, organized canned goods on shelves, scrawled pages in our diaries, or rousted ourselves for a wood-gathering expedition down the beach.

After lunch that day, we gathered up our laundry and hauled it, along with two five-gallon buckets and a plastic fish-cleaning basket, to the pond behind camp. We dipped water, tea-colored from ground tannins and warm from the sun, into the buckets. As our clothes soaked, Molly Lou read to me. She sat on a log over the pond, her toes dipping in the water while I poked at the laundry with a stick. She read a short story by Mary Lavin. *She was islanded by fields, the heavy grass washing about the house, and the cattle wading in it as in water. Even their gentle stirrings were a loss when they moved away at evening to the shelter of the woods.*

As my hands sloshed the clothes up and down, I saw the house in my mind, a dark-haired woman opening the door, stepping onto the porch to look toward something in the distance. I draped the sopping shirts, jeans, and underclothes across alder branches overhanging the pond, my bare feet sinking into the warm mud. When I was done, the tree looked like a rookery. The undersides of leaves flickered with pebbles of light reflected up from the pond. Molly Lou's voice mingled with the flapping of clothes and the whistle and buzz of fox sparrows.

I sat on an overturned bucket. The moment was an anchor line holding us against a strong tide. *This is what we're like when no one's watching,* I thought suddenly, and then someone was watching and the spell was broken. But I knew I would remember it. I imagined a stranger walking on the beach on a similar day years from now, how he'd think he heard voices, how he'd head toward them, and at the pond edge, find bare footprints in the mud, slowly filling with water.

After straightening out the cache, I leave the camp, climbing a deer path into the woods. The sound of the waves abruptly stops. On top of a rise is an old hemlock tree. My arms can only reach partway around it. Its branches weave an impenetrable net, and even during the heaviest rains, I've sat beneath it and stayed dry. It's just a minute from camp but utterly private. Once, from up here, I heard my friend Olga in the wall tent below, singing her children to sleep. Nothing leafy grows in this shade. At the tree's base, visitors leave shells, feathers, blueberries, and cookies. Someone left her dead dog's collar and kerchief under a root. Once, a child left an egg. At eighteen, she came and left a letter. *Dear King Tree* . . . The remains of a wreath Molly Lou and I made last summer dangle from a broken branch.

I lie on the moss. I think I could die without fear on the island. There is some ineffable way the island is—even without my associations with it, even without the camp—that says I'm known, that I belong somewhere.

One spring, after reading a new guidebook to wild plants, complete with recipes and lore for everything from reindeer lichen to devil's club, Molly Lou and I spent our weather days hunting for edible plants. During a three-day storm, we hauled around identification guides, backpacks, and a canvas water bag. Our clothes soaked with sweat beneath our heavy raingear as we searched for fiddleheads, sea lettuce, lamb's quarters, chocolate lily. We'd read about wild crab apple trees and found one above a cove where goose tongue grew in profusion.

Back at camp, we lit a fire and hung baskets of seaweed to dry from the tent's ridgepole until the whole place smelled like low tide. As we walked around, kelp caught at our hair. The storm strained the wall tent's canvas, but inside a halo of heat hovered above waist level. Molly Lou and I dragged beach chairs so close to the stove that our shins burned. Driftwood leaned against it to dry, sometimes blackening where it touched the metal, smoking when it got too hot.

We didn't hear our visitors until they hallooed us from the beach. I made a slit in the canvas doors and peered out. Two rain-suited men and a woman trudged toward the tent, dragging their inflatable boat. I turned and looked at Molly Lou, at the interior of the tent, festooned with dripping kelp, and back to the visitors and hallooed back. "Can we come in?" they asked. "Sure," I answered.

They told us they were a crew of Fish and Wildlife employees, weathered out by the storm. Sitting in a circle, sipping tea and telling each other stories of our research work, I felt giddy from the sudden company. The crew's leader, a fiftyish, thick-torsoed man with a brush-cut, interrogated us: "You mean to tell me that you two live out here by yourselves for four months straight without power? Tell your boss you need a generator. How do you run your computers? How do you charge your batteries?"

"We don't really need a computer out here. We just use field notebooks and data sheets and enter the data in a computer when we get back to town. And I'm trying some solar battery chargers, but that's been a little sketchy," I said. "We kind of like it like this."

"To each his own, I guess," he said.

We knew we were "bushy." Friends visiting us in the Sound often raised their eyebrows at the rituals and language we invented. With our woodstove cranked up, our tent was "dry as a cinder." We called our camp "cawmp," the way I pronounced it as a child. We called one another "Jez." Montague Strait—a dicey passage open to the Gulf—was "Lake La Barge." Instead of storing things, we "stookered" them away. Instead of fishing, we caught "spare fish" by trailing longline boats in the *Whale 1*, netting up their discarded rockfish for supper.

That evening, after the visitors left, we climbed the bluff above the camp. Only wind-deformed and stunted trees grow there. In the peak of the storm, we could barely stand upright. Below us, foam streaked the water's surface, and gulls and eagles rode the updrafts. Raindrops stung our faces and eyes so sharply, we had to look away. In and out of fog sheets, the Pleiades levitated.

Molly Lou and I raced across the bluff top. We found the two tallest trees—not more than eight feet high—and climbed into them, facing the wind. We closed our eyes and felt the trunks shake. Hatless, we jumped off the edge of a rise with our arms outstretched, to see if the wind would blow us back.

Sometimes at night, we'd be roused from reading by a subliminal hum and would look out of the tent flaps to see a cruise ship passing. Strangely silent, it dwarfed the passage, its hundreds of windows blazing. We wondered if there might be a person standing on the deck, looking out at our island and seeing the lit tent, imagining us, feeling an ache of longing in his belly, as if glancing into a warm window in a strange city. We felt lucky.

One summer solstice, Molly Lou and I set out at eleven after dousing our campfire. We followed the trail behind the camp to look for the makings of a wreath. I'd stuffed jingle shells in my pockets before we left the beach. As we climbed from the blueberry woods to the open muskegs, we picked crowberries, shooting stars, ferns, violets, hemlock, and elderberry. I gathered some moss and heart-shaped *Fauria* leaves.

As we wandered between ponds, I thought, and then was certain, I heard voices. We stopped to listen. With a sound like fingernails brushing canvas, the day-breeze still cut across the island from the southwest, pushing my hair up off my forehead. Then—was that a shout from Goose Tongue Cove? I watched Molly Lou bend to pick a flower, motionless with her hand on the stem, her head tilted. When we climbed a hill above the cove and peered down, we saw nothing but waves breaking and alders whipping their skirts.

Someone told me once that the site of our tent platforms had been a summer fish camp for the women and children of Old Chenega. At the back of the pond, timbers rot among wild iris and dwarf dogwood. But whatever those shouts were, they were as indifferent to us as the animals who lived on the island, as the black bear Molly and I had surprised earlier that summer while hiking. We'd rounded a boulder at a pond's edge, and in a blueberry thicket, thirty feet away, a bear sat, pulling branches through its jaws to strip off berries. Remembering the prescribed response, I'd waved my arms over my head and yelled, "Hallo!" The bear had looked up, its black eyes focused on something in front of us that we couldn't see. After a few seconds, it grabbed another branch in a paw and dragged it through its teeth, chewing. We'd backed away, but the bear never looked up again.

After we returned to the beach, we found a long bull kelp whip, twisted it into a ring, and into the loops and curlicues stuck found things: shells, moss, flowers, twigs. We hung the wreath from a nail and stood back to look. It seemed like a window into and out from the island. Where it hung, it framed a view of the passage. We peered through it at the Pleiades Islands hovering just above the water's surface, backlit by an afterglow flaring behind the highest peak of Chenega Island. Molly Lou and I took turns standing in the ryegrass in front of the tent so that we could look at each other through the wreath. "That's the way the island sees us," she said.

At 1:00 a.m., we went into the tent, built a fire, and made tapioca pudding with canned pears. We lit a Coleman lantern, and as Molly Lou told me the story of Ali Baba and the forty thieves, I imagined our world as it might appear from the water, the canvas tent glowing like a lantern mantle, our shadows stretching the walls. Curled up like owls, we drank tea from tin cups. Molly Lou's pupils dilated to black, lights blinking in their centers. Her voice changed. I imagined what we said was heard outside, by the island's other inhabitants.

Once Molly Lou told me that eighteen ravens had surrounded and watched her as she sat in the muskeg. "Surely a sign," I wrote in my diary.

The lives of the island's animals went on without us, and we found traces: bear scat filled with seeds and fawn hooves, rib bones of deer, crow feathers.

It's been hours since I left the boat. I put my diary away and look around. In late sunlight, motes and spiderwebs drift. Soon I should anchor the *Whale 2* in the cove for the night. I take one last walk down the beach, then turn up the trail to the muskegs and pond. I keep looking over my shoulder. I feel completely alone, and I also feel watched. I clap my hands shyly and call out, *Hallo bear* . . . Every leaf is sharp, slightly magnified, and the wind sounds like little feet running.

The next evening, I climb the bluff above camp to scan for whales. From here I can see the mountains encircling the embayment that holds the Pleiades at its center. A wedge of the Chenega Glacier, fifteen miles away, barely visible, sends out rumbles. The sun's afterglow ripples from Chenega to the edge of trees below the bluff, which waver in brightness. Now island shadows lie across the water, except for the Pleiades, which seem to have no shadows. Their marker light has not come on yet. That light intrigued Molly Lou and me, how it required a certain amount of darkness to trigger its work.

Suddenly, the island's awake with birds calling. Guitar-shaped ponds gleam like metal. Seabirds bleat on the northern point. It seems the darkness begins on the island, between trees. I look at the view for a long time. I try to widen my eyes.

I think of Old Chenega, how, when the ceiling drops, icebergs drift by the islands in front of the village, their blue intensified. Mist fingers down from the mountain behind the village, grazing the ruins and crosses. Owls flit among branches. I saw them, once. Years ago, a young Chugachmiut woman took me for a walk around the old village. Near a collapsed house overgrown with blueberries, three owls darted among spruce saplings within an arm's length of my face. I imagine them now, watching over the old village, their eyes glinting.

The island's not mine. I like to think that we created our life there, Molly Lou and me, with our talk, our language, our walks and gathering, but I think all along, it was the island creating us.

Sometimes, when I can't sleep, I imagine the island in winter. The hematite ocean runs high above tide line, right at the edge of the tent platforms. Huge logs roll around in the water, logs that once, in summer, seemed sunk into the beach, solid as pilings. The waves are so high, the beach seems below sea level. Wind and sleet belt the trees, swaying them crazily. I imagine deep snow down to the water, the tent platforms bowed under its weight. When deer descend from the high country to graze the shoreline for kelp, their coats are dark with wet. They hold themselves against the gale, which sweeps the island clean of ghosts, shaking it in its huge hands.

©BECKER '00

Crossing the Entrance

Let little boats brave the waves, and at dusk
The spiky anchor light of one green ship,
A kind of comfort, a kindness
To the beasts of the sea and the wood.

— Molly Lou Freeman

In quiet anchorages, in familiar surroundings, ground tackle and the methods used are seldom put to test. Cruising into strange waters, finding inadequate shelter in an exposed anchorage during a hard blow, and the elements will surely take the measure of both tackle and technique. . . . Ask yourself now, in a situation like that, would I hold . . . or hope?

— *Chapman's Piloting*

I wake suddenly and place myself: curled in my sleeping bag on *Whale 2*, moored in a quiet corner of Gibbon Anchorage. Then I hear it again, scratching through the VHF radio: *Mayday. Mayday. Mayday. This is the fishing vessel* West Wind *in distress 2.5 miles south of Johnstone Point in Hinchinbrook Entrance. We're a hundred-foot vessel with four persons aboard. We're in survival suits, and we've activated our EPIRB. Mayday. Mayday. Mayday.*

The message is repeated, Loran coordinates given. A solemn-voiced Coastie replies instantly. *Fishing vessel* West Wind. *Fishing vessel* West Wind. *This is Air Station Kodiak. Channel one six, over.* Long silence. The Coast Guard hails again. There's no reply from the *West Wind*.

A sob rises into my throat as I imagine the boat foundering in the Entrance. Far offshore, four people huddle in survival suits, pulled by currents, washed by waves.

Other voices break the radio static, boats traversing the Entrance who'll adjust their courses and look for the distressed *West Wind.* The Coast Guard warns them to beware of floating debris. I imagine that flotsam, the buckets, fishing nets, life rings, buoys, planks, clothing, sodden charts, shoes. A diesel sheen spreads from the wreck. A tiny swirl marks the spot where the boat went down.

Fifteen minutes later, the Coast Guard dispatches a helicopter. After another hour, they call off the search. *Pon-pon . . . pon-pon . . . pon-pon . . . All four crew members of the fishing vessel* West Wind *have been rescued by Coast Guard helicopter and are being taken to Cordova Community Hospital.* This is all I know of the *West Wind*'s fate.

I've stared at nautical charts, read the fine, italicized print. *Caution . . . Heavy Tide Rips . . . Unusual currents may be encountered . . . Mariners are urged to navigate the area with caution . . .* Inside Hinchinbrook Entrance is Prince William Sound, sheltered from the ocean by the two big islands. Outside is the Gulf of Alaska, where some of the stormiest weather on earth is. To the east, past Kayak Island, the coastline is known as "coffin corner." To the west, on the nautical chart, black isosceles triangles—symbols for shipwrecks—dot Montague's outer coast.

I listen intently when mariners speak of the Entrance. I've a strange penchant for seafaring stories of storms, ship disasters, and big waves, considering that I'm an anxious person. For the past several summers, I've lived on the sea for four months at a time, often as the "skipper" of the boat, but new experiences and unknowns still scare me. In my body, I'm a thirty-four-year-old woman, and today she has to take a boat across Hinchinbrook Entrance. But this body houses the same self who, at seven, stood shivering, clutching her sides, bony knees pressed together on a diving board's edge while a swim coach cajoled from the water below, and other children catcalled from behind: *Jump!* And I didn't.

People say I seem calm. But along with my research assistant and me, there's a third being on the boat, invisible, wringing her hands, chewing her fingernails, fretting. She's been my companion all of my life.

We take a compass bearing, west. The chart lies open on the deck. It's five in the evening, and my field partner and I have fifty miles to cross over Orca Bay and Hinchinbrook Entrance to Green Island, where we've stored fuel for our summer's research. The sun won't set until eleven o'clock, but there aren't any nearby islands to serve as landmarks. What looks like a clump of cumulous clouds low on the horizon is Montague Island, we guess.

Entering open water, we pass between the flat face of Red Head on our right and Gravina Rocks on the left. I hug the left-hand shore to avoid a three-fathom bank. Drawing only a few feet of water, our boat could pass over the bank without our even knowing it, but the pale blue amorphous shape marking the shallower water on the chart keeps me away. On a chart, blue marks water less than eight fathoms—forty-eight feet—deep. Most of the Sound is white, with canyons a thousand feet deep, gouged out during the glacial age.

Whale 2 is an ex-fishing boat with a structurally questionable plywood cabin mounted on the stern and a long, open deck where the gill-net reel once perched. Built in the 1960s to fish the Copper River Flats, her hull's seaworthy, but she shows hard use and idiosyncratic modifications of the previous owner, who called her *Bonnie Blue* and mounted all of her windows upside-down, so they leak. Baking pans catch the drips. The cabin is tiny, bathroom-sized, with a diesel stove for cooking and heating and a foam mattress laid across the engine compartment for sleeping. Despite her Alaskan funkiness, she's solid, built for these waters.

I don't feel seaworthy enough to match her. It's my first time using *Whale 2*, my first time piloting a boat across the Entrance, my first time working with Laurie, an experienced field biologist who's tracked radio-collared wolves from small planes through the Alaska Range but who's never worked on a boat before.

The water's calmed since Laurie and I first tried to cross the Entrance earlier this morning. We left Cordova Harbor at seven, hoping to avoid the southwest day-breeze. Fifteen- or twenty-knot winds, beginning midday, build three- to five-foot white-capped waves in wider bays and passages on sunny days. I didn't want to be caught in Orca Bay, with its long fetch—twenty miles of unsheltered water in any direction—in a strong day-breeze, especially with our recent fuel problems.

We'd been towed to Cordova three days earlier after taking on watery diesel and damaging the engine's fuel injection system. It was embarrassing. At the first sign that the engine was running roughly, I hadn't paid attention. I'd assumed that diesel engines worked similarly to the outboards I'd used in the past, and my repair attempts had driven water deeper into the fuel pump's intricate coils.

Ed, the mechanic in Cordova, had loaned us a Sawzall and an electric pump to cut open the fiberglass fuel tanks and remove gallons of watery diesel. While Ed replaced fuel injectors, Laurie and I walked around Cordova. I felt like a failure, especially when I called our boss, not with news of whale encounters, but with a warning that a thousand-dollar repair bill was in the mail. At the bookstore, I bought the sixty-first edition of *Chapman's Piloting*, a tome of seafaring, and *How to Become Your Own Diesel Mechanic!* I vowed to read them every night.

I'd also bought a card, which I taped up in *Whale 2*'s cabin. Underneath the Chinese character for new beginnings, *ch'un*, was written:

> *Times of birth and growth start unseen, below the surface. Everything is dark and still unformed, yet teeming with motion. Difficulties and chaos loom. Despite this struggle, energy and resources are collected and form begins to take shape. The young plant takes root, rises above the ground and is brought to light.*

That's what I wanted, new beginnings. I resolved to be vigilant with the boat. After leaving Cordova, we stopped every hour to check the filter for water. I marked our progress in the logbook:

9 July

0700 leave Cdv sunny B1V2 < 5 knots wind
0750 check Raccor 3–4 Tbsp H_2O in filter
0850 1/8–1/4 cup water
0950 3 Tbsp
1000 Abeam Middle Ground Shoal; B2V2; wind SW 10 knots
1050 Abeam Johnstone Point; wind West < 10 knots; 2 Tbsp H_2O

Then we spotted glinty black wedges in the Entrance, a pod of thirty killer whales moving toward us, wind blowing their spouts along the water's surface. We reversed course and paralleled them. While Laurie maneuvered the boat, I snapped identification photographs of each individual for our census, matching the whales' fin shapes to pictures in our catalog, jotting notes in the field book:

1117 AB and AI pods traveling, dispersed/subgroups 1 mi N Johnstone Point
1313 bet Gravina Is and Johnstone Pt whales seem to be feeding here
1334 Males play and some feeding

1350 whales spread out and feed—silver salmon?
1355 B3 V2 west wind 10–12 knots

During a normal encounter, we might spend twelve hours waiting for perfect light for photography, taking hundreds of pictures to make sure we didn't miss new calves or shy animals, noting changes in behaviors, trying to collect scales from fish kills, recording calls with a hydrophone, but I was uneasy. The whales were taking us off course, and the day-breeze was rising.

We anchored in Port Gravina to wait out the seas. As swells lifted and lowered the boat, and beach stones clattered and hissed in the surge, we dozed on deck until the water calmed.

To enable the helmsman to steer a compass course, a lubber's line is marked on the inside of the compass bowl to indicate the direction of the vessel's bow.

—*Chapman's Piloting*

Until we make out Montague Point, we'll hold our course, splitting the duties, Laurie at the inside helm, watching the compass and gauges, me steering from outside, watching for flotsam. Eelgrass and kelp could clog our intake, causing an overheat, while a log might bend the propeller. More than that, I want to see distant shorelines resolving into something familiar, the surface of the water like a map, with dark lines where the wind touches down, and strange, glassy floatings where it's calm. Laurie calls to me, "You're drifting east," and I adjust course.

Fifteen miles out, we stop to check the fuel filter. We try to match the chart to what we see. Through binoculars I scan Hinchinbrook Island's shore until I spot a faint orange dot, the Johnstone Point day marker. To the west, the horizon distorts. A small blue island floats where Montague should be. Where are Montague's snowcapped peaks? What if the compass is wrong?

I recognize the glittery sensation in my body, the voice in my head. The fear is limbic, an icy stream between my body and brain. I've also known it by a different name, what my father called "stage fright." In high school, I wanted to be a musician, but panicked during auditions. An undercurrent tugged at my legs. I made long checklists in my diary, bit my nails until they bled. At fourteen, my hair fell out.

Fear is also my inheritance. While I feared devil possessions, nuclear war, house fires, and airplanes, my father feared high places, edges, mountains, travel, and lightning. During violent, nighttime thunderstorms—frequent summer occurrences on Lake Erie, where I grew up—he thumped through the house, slamming windows, then pulled us from our beds and hid us in the basement. This is memory: at fourteen, in Latvia, he watched a village of thatched houses, lightning-struck, burn. Near eighty, he quadruple locked the doors, wandered at night, an insomniac and chronic worrier. This might also be recollection, a bodily memory of upheaval—like the earth's memory of earthquake or burial—how, at eighteen, he was drafted into the Latvian army, fought on the lines, was taken prisoner, escaped, and never returned to his homeland.

I love thunderstorms and don't lock things. I've climbed mountains, swallowing my fear of edges. But fear is coiled up in my chemistry. A few years ago, on a turbulent airline descent into Anchorage, I clutched the seat in front of me, weeping. The heavyset woman in the next seat tried to calm me, telling me she could see the lights of town, rubbing my shoulders. When the plane slammed down, I sat back and gulped air.

"Thank God," I sighed.

"Honey," the woman said, "you've got to learn to relax!" I looked at her and nodded.

I distrust the compass, so I change course, toward Hinchinbrook Island. We'll follow the coastline to Johnstone Point, then cut straight across. As we start up again, the sun's belly dips and touches the white spine of the Chugach Mountains, a glaciated arc that forms the Sound's northern rim, and a deeper blue creeps across the Sound like fog.

With a strong S gale and ebb tide, very heavy overfalls and tide rips occur in Hinchinbrook Entrance, and are dangerous to small craft. Tremendous seas, steep and breaking, are sometimes encountered just outside the entrance. During heavy weather, tide rips and confused seas are in the vicinity of Wessels Reef. Many halibut schooners have foundered between Cape St. Elias and Montague Island.

—Coast Pilot 9

As we rumble toward the Entrance, a dialogue runs wild in my mind. I should have read a diesel book *before* taking the boat out for the first time. I should have taken a mechanics class. If a storm comes up, can I get us through it? If we overheat, will I know what to do?

Last summer I spent two weeks aboard *Whale 2* with another researcher. She'd taken a diesel class, had several years of experience. We hit rough seas on our way back to town. Waves slammed the hull, splattering on our faces. Salt water streamed across the deck. Between larger waves, *Whale 2* wallowed for moments, then rose again. When my friend asked me to switch fuel tanks, I crawled to keep from falling. In the cabin, drawers had slid out of their slots, throwing tools, silverware, pencils, and spools of thread on the floor, where they rolled about. Books and cameras perched precariously on the shelf. I grabbed them and threw them onto the lower bunk. After turning the fuel valves, I stumbled back on deck, nauseated. Kathy drove calmly, smiling, singing to herself. She steered with one hand, her fingers knotted over the knob bolted to the wheel. The suicide knob, it's called.

I need that kind of confidence now, I think.

> *You will be more confident and your judgment will be better if you can meet the storm as an old and familiar adversary rather than a new and unknown one.*
>
> —*Chapman's Piloting*

Years ago, I crossed the Entrance in darkness. On wheel watch for two hours while everyone slept, I navigated my friend's seiner by green instrument lights of radar, Loran, auto pilot, depth sounder. I studied blips on the screen, translating them into ships' lights and shorelines. The cabin lights were off. Peering through the window, I tried to discern the shore but saw only my wide-eyed face reflected in the green-tinted glare. What if I misunderstood the directions for reading the Loran? What if we hit a rock? My friend dozed in a bunk a few feet away. He trusted me enough to sleep.

My last crossing was a few days ago, under tow to town. A fishing boat heard our radio call for assistance. We'd rigged a line between the two boats.

Laurie and I rode on *Whale 2* in case the tow-rope broke. I sat on deck and wrote in my diary:

> *I feel sick. We are being towed like some unwanted thing behind the* Illaquare. *Like some useless and broken thing. They just threw overboard all of the fish they caught that went bad because they couldn't make it to Icy Bay last night—because of us . . . In years past I've entered the Sound gratefully. Now those times seem distant. The water is wide and vast and turbulent to me (though it's glassy calm here). I'm so engrossed in engine mechanics, I haven't even connected to the place. I feel alone and adrift . . . We are going to town and we're going to have to come back from town, across the Entrance.*

Now the wind smells sharp, like aluminum and kelp, and my hands stiffen on the pot-metal wheel.

> *Johnstone Point, 57 feet above the water, is shown from a skeleton tower with a red and white diamond-shaped day mark on a pillar of rock off the point. The flood entering W of Montague Island sets NE past Montague Point and causes rips between it and Johnstone Point.*
>
> *—Coast Pilot 9*

Rips jostle the water off Johnstone Point, where the navigational light blinks, a guide. The sun's last burning wedge slides into a sharp juncture in the mountains, and the sky bends its back toward the afterglow. Light sinks into the sea and glows up again, like jellyfish shoals, their jade lamps glowing. *There is still quite a bit of light left*, I think.

Once past the point, into the Entrance, confused seas pitch the boat from side to side. Beneath that, ocean swells from the Gulf lift and lower us. Laurie joins me on the deck. "How are you doing?" she asks.

"All right," I say. "We're in the washing machine now. But we're getting there."

You're full of shit. How much longer will this take? What if it gets dark? What if the waves get bigger? My legs shake. It's too dim to tell wave size. They hit the starboard quarter and break into spray. When I'm splashed in the face, salt pricks my skin. Laurie grabs our raingear from garbage cans lashed to the deck. She holds my coat for me as I fumble for the armholes, shoving my arms through stiff sleeves.

"I'm feeling pretty sick," Laurie says. Her face is pale, her hair damp. "There's a foam pad in the garbage can," I tell her. "Lie down on the deck so you can get fresh air."

"Are you sure you're okay?" she asks.

"Fine," I say. I'm grateful not to admit my fear. Laurie curls up on the pad, her body rocking with the boat's motion. She must trust me, I think, and I envy her.

Along the blackness of Hinchinbrook's shoreline, toward the place where the Entrance meets the Gulf, I see ship's lights. They're coming into the Sound, and the Entrance is the beginning of shelter for them. For a while, they're company. There's someone else out here. There's someone to call if we need help.

The boat slams hard. Laurie's body tenses. When I check my watch, it's been forty minutes, though it seems much longer. Montague Point is still miles away.

> *Meet each wave as it comes. You will be able to make reasonable progress by carefully nursing the wheel—spotting the steep-sided combers coming in and varying your course, slowing down or even stopping momentarily for the really big ones.*
>
> —*Chapman's Piloting*

When I check my watch again, it's 11:30 p.m. We've been on the boat since 7:00 a.m. I've completely misjudged the distance across the Entrance. It will be dark by the time we reach Montague Island. My eyes strain to see Rocky Bay's mouth, where we can anchor. I want to shake my fist at the wind, the waves, the distance. I don't know what or who I'm angry at, but each time a large wave jars us, I clench my teeth. Hours ago, we gave up checking the fuel filter, marking our log. Like a child, I pray, *Please God let us cross safely. Please help me find Rocky Bay. Please help me anchor safely.*

And I can't stop the thoughts. How someone once told me about boats with "snowball" hulls, like the *Whale 2* has. In big following seas, they can pitchpole, capsizing end-over-end. I imagine this, what kind of rogue wave might bloom behind us, hesitate like a clawed hand before driving our bow under water. I remember a story I heard, how a bowpicker like *Whale 2* sank after pitchpoling near here. When I asked my boss about it, he exclaimed, *That's not even true! I don't even want to hear about that.* That man, he told me, had been a fool, out in a bad storm with broken steering. Two Coast Guard officers died when their helicopter crashed trying to rescue him. He eventually made it back to town on his own. *People have fished those boats off the flats for years and years,* he said, annoyed at my chronic worries.

> *Excessive speed down a steep slope may cause a boat to pitchpole, that is, drive her head under in the trough, tripping the bow, while the succeeding crest catches the stern and throws her end over end.*
>
> —*Chapman's Piloting*

Back when it was dry-docked in the boatyard, Laurie and I had to make repairs to *Whale 2*'s engine. When I lifted the engine box cover, bilge stench wafted out.

I couldn't decide where to begin. To delay picking up tools, I read repair manuals and cleaned the engine's greasy crevices with rags and a toothbrush. Finally, I tried to install the new alternator, but one of the locking nuts was missing. The new saltwater pump wouldn't fit onto the cooling water tubes.

Exasperated, I took a long run. Several miles down the road, I saw a wooden fishing boat sunk into a field of dead grass, blistered paint peeling from her squat, bowed-out belly. A wheelhouse perched on top, the window glass broken into shards. I imagined the blocks she rested on pressing deep into the soil, the hull's waterlogged planks soft enough to put my fist through. Passing her, I heard a rattle, and I stopped. The slotted aluminum cap on the stovepipe spun in the breeze. I saw myself looking down at the *Whale 2*'s engine, brain whirling, body immobile.

When I returned from my run, I asked Jim, the Australian mechanic, about the saltwater pump. "Just jam it on there, love. You won't break it. It's supposed to be a tight fit. Just make sure you check it for leaks when you start it up."

As Jim gave us a quick rundown on diesel operation, I watched his grease-imbedded hands deftly hook a water hose to the saltwater pump and shine a flashlight around the engine room, his voice twanging out instructions. "Start her up, Evie." The engine roared, shaking the boat on its blocks. Jim peered around, grinned. "No worries, love," he said. "Just carry a spare impeller, keep your fuel clean, and you should be fine."

Outside, I watched him light a cigarette and shuffle off in oily jeans to the next person in line. That night, I dreamed I was going on a rocket to the moon. I was terrified until I realized that a biologist I knew, a man, was going with me. The next morning, Laurie and I launched the boat.

~

As we approach Rocky Bay, the wind rises, cold, insistent, blowing from the bay's head. I don't look forward to a night of worrying if the anchor will hold. "We're almost there, Laurie," I say. "Are you all right?"

After a pause, she replies, "I'm better." I turn on a flashlight to look at a chart of Rocky Bay, but the light blinds me and I switch it off.

~

> *Rocky Bay is deep, and exposed to N and E winds. A small vessel can anchor in good weather about 0.6 mile from the head and 0.2 mile from the NW side, in 5 to 6½ fathoms. Small craft can anchor in the lagoon, on the S side 1 mile from the head, where a small area has a depth of 10 feet. When entering the lagoon care should be taken to avoid a reef, partly bare at low water, extending W and NW from the N point. A dangerous sunken wreck is in the entrance to the lagoon.*
>
> —*Coast Pilot 9*

After midnight, we enter the bay, guided by murky edges where rocks rise out of the water. There's no light left. I slowly ease up on the throttle. In darkness, it feels as though we're gliding faster than we are. If we go farther, we may hit submerged rocks or ground on the mudflat in the outgoing tide. Laurie rises, stumbling. She unlashes the anchor, rests it on the bow while I shift the boat into reverse, backing toward shore, so the anchor will catch on the rising bottom. "Let her go," I call.

When I hear the anchor splash, I imagine it fanning down like a flounder. It strikes the bottom of Rocky Bay, catches its fins in the gray and slippery mud, and holds.

Before going inside, I collect my flashlight, binoculars, and log. I open the soaked pages to where I left off. My fingers are so stiff, I can barely grasp the pencil. I scrawl:

0100 Anchor in Rocky Bay.

The bitter end of a line's the extreme end, the end made fast when all line's been paid out.

—*Chapman's Piloting*

We strip off our damp clothes. After picking up books and cups from the floor, I crawl into the bunk and into my sleeping bag to warm up while Laurie makes miso soup, popcorn, and tea. The stovepipe's cap whirls in gusts. After dinner, I go out on deck to check the anchor.

The wind is sharp, like a hand against my face. I choose a prominent, leaning tree on one shore and a small island on the other as markers. The anchor's holding. The wind's steady, but not really that strong. I pick up an oar and stir up a cloud of green phosphorescence. Below, I imagine sunken ships planet-like and still. Behind me, light shines from the cabin windows. How safe we'd look from a passing boat.

The next day, we begin our three-month field season. At night, I read the engine manuals. Every morning, Laurie and I run through a troubleshooting list before starting up: check oil, coolant level, belt tightness, fuel filter, saltwater strainer. We label a field notebook "Engine Log" and record engine hours and maintenance.

A few weeks later, in Lucky Bay, Laurie makes breakfast while I check the engine. When I turn the key, the engine wheezes once, then stops. My stomach feels like a stone's dropped through it. "What's wrong?" asks Laurie.

"I don't know. I think the batteries are dead." I turn the key again. Click. Nothing.

"We'll have to call the fish hatchery on the radio, and if we can reach someone, see if they have a charger," I say, dismayed. Dead batteries are a sign of careless seamanship, I know.

Another research boat in the area responds to our call, and to our relief, says they'll come right away.

After their boat's tied to ours, three men watch as one hands jumper cables into *Whale 2*'s cabin window and tells me how to hook them to the battery. When I turn the key, the engine starts. But why did the battery go dead? It's rare to have another boat around here. Next time we might not be so lucky. A man comes into the cabin.

"Did you leave lights or equipment on by mistake?" he shouts over the engine's roar.

"I don't think so."

"Well, it might be your alternator." He leans over the engine control panel to look at the gauges. "Look, the needle's just at 12 volts. Your alternator's not charging."

"I thought it was supposed to read 12 volts," I say, puzzled.

"Oh no, when it's charging, the needle should be above 13."

Why didn't I know that?

The man shuts down the engine. "Let's check your connections real quick." Laurie digs in the tool drawer for wrenches. Not only are the alternator connections loose, but the wing nuts holding the battery cables in place are loose too.

"You've got to tighten these down," the man says. "Vibration wiggles them free. A loose connection can cause electrical arcing. A boat at the hatchery caught fire that way."

As their boat pulls away, I imagine their conversation, how appalled they must be at our ignorance.

"Eva," Laurie says, "why don't you row to shore and take a break. I'll make lunch. Then we'll leave." Laurie sees our mishaps as part of a natural learning curve. But she also speculates on the deeper purposes of difficulty. She's brought a packet of small cards with her. Each card's imprinted with a single word—harmony, synchronicity, honesty, fate, knowledge, serendipity—dozens of words, and every morning, we close our eyes and meditate on a question, then draw a card. We look up our words in the dictionary, try to divine their true meanings. What does it mean that the engine keeps breaking down, I ask. What does it mean that we can't find whales? Why am I so afraid of this boat?

Two streams emerge from the forest and fan across an intertidal flat before meeting the green of Lucky Bay. The air here smells of wet spruce and kelp. In the middle of a stone circle, a small fire burns. On one of the stones, an empty mussel shell, its two halves held together by dried sinew, opens like an entrance. Kelp hisses, drying above tide line.

On paper scraps, I scribble words: *Bad Luck . . . Self-Judgment . . . Pessimism—always thinking of the worst that could happen . . . Stress, Worry . . . Seeing breakdowns or other mishaps as somehow reflecting my place in the world . . .*

I drop them into the fire.

A puffin flies across the bay, showing me that the water's not so vast or dangerous, but a smooth landing place, a negotiable size.

Once, my mother, sister, brother, and twelve-year-old niece came to Alaska. They came without my father, so we climbed a mountain. Halfway up, Emily, my niece, spotted a wave-shaped rock and wanted to climb it. My mother sat at the rock's base while Emily led the rest of us through a snow swale, up a scree slope, to the rock's crest. Emily and I reached the top first, and I crouched, looking into the canyon below. Exuberant, Emily stood and waved to my mother. I fought the impulse to tell her to sit down. My mother called out, but the wind swept her words away. When I looked back, I saw my sister, crouched several feet below, and my brother, hunkered in a similar pose, farther down. I looked beyond them to the tangled alder thickets sloping to the Delta River's gravel bed, a thousand feet down, twisting and braiding, cutting new channels and banks when it outran the old. I saw my father's figure in the far distance. I saw the unraveling rope of inheritance, hung with pitons, knots, and carabiners, joining my brother, sister, and me to him. When I looked at Emily, I saw the space between us where the rope was undone.

If fear is somatic, its only antidote is the body, touching the earth, repeatedly, in new ways, teaching the hands, eyes, torso, pelvis, thighs, and face a new language. As Laurie and I encode our hands with memory of tools, as we learn the shape and heft of stainless steel, the click of a socket onto a ratchet, the sudden loosening of a frozen carriage bolt, it's like we're writing a new engine mechanic's guide. It's illustrated with photographs of Laurie's brown, sturdy hands holding tools, of my small, wrinkled hands grasping wrenches or gauging the tightness of alternator belts. Oil works its way under our fingernails. In our engine manual, all the pronouns are "she." Fingerprints stain our log's pages:

26 July 93

0700 Leave Lucky Bay; B1V1, calm and lovely and smelling of the open sea
1030 1 mile NE of Crafton Isl. light. Hear killer whale vocalizations faintly, getting louder B1V1, calm, some variable breezes, a few high clouds

One-Hundred-Hour Maintenance

No one who survives to speak
new language, has avoided this:
the cutting-away of an old force that held her
rooted to an old ground

— Adrienne Rich

Kathy and I prepare to drop anchor. Except for the cabin light of a sailboat flickering on the water's surface, the cove is dark. I push the shift lever into reverse to ease *Whale 2* toward shore. When I shift back to neutral, nothing happens. We're stuck in reverse. I turn the wheel hard over.

This has happened before, I tell Kathy. I rush inside and shut off the engine, and the boat gradually slows its drift.

While Kathy shoves the anchor overboard, then holds the line, feeling for the anchor's catch, I unscrew the shift lever mechanism inside the boat. I lift the metal box, peer at the cable clamps, but everything's in place. Grabbing a flashlight, I jump outside and lie on deck, looking up at the inner workings of the outside steering station. The clamps and cables look fine. But the shift lever flops back and forth with no tension at all.

"The anchor caught," Kathy says. She's looping the line around a cleat. "It's late, Eva. Let's get some sleep and deal with this in the morning."

I'm exhausted. It's near the end of a three-month-long field season. We searched twelve hours today without finding killer whales. My eyes ache from staring through binoculars at sun glare on water. But I can't help myself. Inside the boat's cabin, I open the engine cover and inspect the shift cable, where it disappears into the stern, heading toward the transmission on the outside of the boat.

As I pile engine manuals on the small table by the bed, my mind moves jaggedly over what I know and don't know, a pencil held in a child's fist, veering off the page. While Kathy nestles into her sleeping bag, I leaf through photographs of gear cases, search for a troubleshooting chart. When I switch off the light, I hear Kathy's slow breaths beside me. I try to match them.

I met Kathy at the university in Fairbanks, where we were both graduate students. I joined her yoga and T'ai Chi classes. In the marine biology department, she studied walrus thermoregulation from data she collected on cold, windy islands in the Bering Sea. The same large, sea-colored eyes that observed walrus skin blooming or fading with heat watched the body moves of T'ai Chi students, noticed chickadees feeding on grass seeds outside her window. Now, she's training herself to listen, recording natural sounds on digital tapes—the clicks, grunts, and thunder of caribou herds, the night sounds of boreal forest birds, creek music, voices of Alaska Native people—and producing stories for radio. She's come along on the boat to help me collect data for the killer whale census and to record sounds for a radio piece. For the last two weeks, we've taped sea lions groaning on the Needle, rain dripping in a hemlock forest, and at 3:00 a.m., a sea otter mother and pup, tussling and squealing. We've followed killer whales. We haven't had a single mechanical problem.

Sometimes, after a long day's running, I feel the engine's warmth beneath me as I lie on the bunk, and it's a comfort. But tonight we're sixty miles from a mechanic, broken down, and the heat is a monster's diesel-tainted breath.

T'ai Chi evolved from Chinese martial arts three centuries ago into a set of postures performed in time. Subtle gestures transform one posture into another. Palms cutting themselves, with a shift of hand, foot, and face, become a white crane spreading her wings. A platter between the hands becomes a cloud.

Before beginning the form, I prepare. My arms hang loosely at my sides, my fingers relaxed. I place my feet hip-width apart and feel my weight on my heels. I relax my shoulders. My eyes, soft focused, look straight ahead. I think nothing. I begin.

At first light, I row the dinghy a quarter mile to the beach, where some friends are camped. I borrow their skiff to get to a spot with good radio reception to patch a call to a boat shop on the marine operator channel. Ed, who's been a mechanic for forty years, instructs me in his gruff voice to beach the boat, open up the transmission, perform two tests to confirm if it's the shift cable or the gears. I imagine how my words translate as schematics in his mind. I thank Ed, but all he says, as always, is "Okay."

Kathy and I beach the boat, stern first, on the outgoing tide. We have to wait for low tide before we can work on the transmission.

"Eva, why don't you just forget this for a while. We'll fix it."

"I can't *not* think about this boat, Kath," I shoot back, my hackles rising in defense. And it's true, even in sleep, I'm listening for bumps, laps, scrapes, wind gusts. "If I stopped thinking about it, and something went wrong, the boat could sink."

She rolls her eyes. "If you can't stop thinking about this boat for ten minutes a day, then there *is* something wrong. Use your meditation practice. I know you want peace."

I want peace. But my mind is like the *Whale 2*'s engine room. While the surface looks clean and well-cared for, in the joints and crevices and floating around in the bilge is muck. When I reach into its cold slop, I never know what I might find: lost screws and washers, broken light bulbs, globs of congealed grease, cracked oboe reeds.

In memory, I watch for it. The moment when the slim white plastic baton flies from his hand. It bounces high off the music stand of the kid who played a wrong note, of the girl who talked during a rest. I'm not the one, so I relax. "Jesus Christ," he shouts, striding to retrieve his weapon. I want to say, *You're an atheist. Why do you shout at God so much?* But I don't, because I'm afraid of the flying baton. And I love his passion. Music is his God. He conducts us like a fire and brimstone Baptist minister.

He stops the band in mid-stride. Points to the flute players sitting in front and demands each one play a difficult section, singling out the ones who haven't practiced. Their cheeks redden from the effort. Sometimes someone cries. Most don't give a damn. For them, band is an easy credit. I can't understand that. For me, it's the hardest credit of all. I attach myself to my oboe as I would to a lifeboat. The flying baton might puncture an air chamber, and I would sink.

Tchaikovsky said, *Truly there would be reason to go mad were it not for music.*

The form begins. Once I begin, I'm committed.

Shift weight left, sink right. Lift palms slowly up, then push down the waterfall. Extend left leg, foot flexed, toes pointed to the sky. Lift hands away from heart.

At anchor, the boat floats again. Following Ed's instructions, we found the broken shift cable. The new one will be on tomorrow's mail plane, which will deliver it to a hatchery fifteen miles away. We'll borrow our friend's skiff to go and get it.

At dusk, I walk down the beach alone. On the northernmost Pleiades Island, the marker beacon blinks. I find a log and sit on the stones with my back against the wood, listening to the breeze rattling alder leaves behind me, the rocks gulping waves. When I glance over my shoulder, I see the stump of a worn-down branch squatting there, a good-humored Buddha, watching me.

I close my eyes and breathe in. At the base of my belly, something trembles.

Oboe players are crazy, Mr. K—— teased me. *They go crazy from blowing all that air through that tiny reed. They blow their brains out.* I thought it was the other way around. Inside me, words emulsified. If I didn't transform that thickness into music, I'd go crazy.

Now, silence undoes my thoughts in the moment between breathing in and breathing out. Breath pushes the trembling to the edges, and, like water, envelops the anxiety encrusted there, and it begins to dissolve.

> *Crossing palms, sink onto right leg. Eyes look to the right diagonal. Extend hands softly to right diagonal, reaching for a ball. Pull ball to left hip. Offer ball to keeper of the east. Pull his arm to left hip, then carry him on your heel, and push him away.*

Mr. K—— admired my parents. My sister and I wore homemade dresses to school. We didn't talk back. We practiced our music, took honors courses, studied Latvian in correspondence school. "Your parents teach you respect," he said, handing me a pink slip—a pass to leave study hall to practice oboe—"you have responsibilities at home." When I took the slip, he tugged back on it. *Vicarious hand-holding*, he called it. In the blank space for my name, he'd scrawled "Boobs." Even now I feel heat in my face. He laughed, wrote out another pass.

> *Right hand becomes the sun. Circle the sun with left palm. Twist torso to the east, right hand forming soft fist. Punch the panda's belly. Unwrap the sword.*

My parents taught me about silence. They rarely talked about their lives in Latvia or about the war that ended their childhoods. I knew they'd met at a displaced persons' camp in Germany after World War II. Latvia had been annexed to the Soviet Union. Afraid to return home, they sought permission to emigrate. At extended family gatherings at my aunt and uncle's house in Toronto, in the dimly lit atmosphere of alcohol and cigarette smoke, I hovered along the table edge, listening, smelling the yeast and blood aroma of saffron bread, head cheese, smoked eels. In a back room, men puffed cigarettes and played cards, whiskey glasses ringing the table. In her tiny bedroom, my grandmother, ninety years old, rocked, fingering her wooden rosary. When she saw me in the doorway, she stopped, beckoned, offered me a tiny glass of thick, sweet liquor.

At home, my father punished our misdeeds with his hands or his belt but mostly with his silence. By way of explanation, my oldest brother once said that my father had a personal hell. I imagined him in his basement office with its walls of books and old maps, its mildew smell, his memories smoldering around him: himself at fourteen, a peasant boy in Latvia, alone with his father when he died of a stroke; at seventeen, pistol-whipped by a German commandant; at nineteen, fighting on the frontline, starving, legs frozen, captured by the Russians. When he's thinking, his lips pull back at the corners in a grimace, his top teeth biting his bottom lip. When I asked my mother why he made that face, she said, "It's because he's in pain."

Naturally, melancholy music attracted me: Solveig's song, Bach cantatas in minor keys, Shubert, Tchaikovsky, Brahms. To reach him in his lonely place, I played sad music for my father to hear.

Pivot to southwest, from left to right leg, turning halfway around the world. Hold water near belly with cupped palms. Pull right knee to chest. Balance ball on toe, reaching out to touch the sky.

I want to know everything about engines now. But any mechanic will tell you the secret is maintenance. We learn kinds of maintenance through experience. What sounds to listen for. What tools to have on hand. When to act. When to employ the old adage, "If it ain't broke, don't fix it."

After we resume our research routine, I teach Kathy daily- and hundred-hour maintenance. Every day, we run a flashlight beam over the engine's peeling green surfaces, looking for drips and leaks. Our fingers press belts, twist wing nuts. We squeeze our bodies into the engine box, wriggle our hands into greasy crevices. Every few weeks, we read off the items on the hundred-hour checklist: change filters and oil, check battery electrolyte level, check oil level and condition in lower unit, grease upper and lower steering shaft journals.

Kathy teaches me another kind of maintenance, how to slow down, how to sink into the earth through my feet, how to balance on one bent leg. That kind of slowness requires strength; it is the heart of T'ai Chi. But I'm not naturally good at slow. If I can't run five miles, I won't run at all. "Run for twenty minutes," Kathy says. "Take ten breaths."

"Fake it 'til you make it," she tells her T'ai Chi students.

> *Leading with the left toe, form a crane's beak with right hand. Hide the beak with left palm. Step out with left toe and open to the west: white crane spreads its wings to push the air away.*

Prepare. Imagining the jobs ahead of me, I reach into the tool box and pull out things I need, lay them on pieces of paper towel. From the upper shelf above the bunk, I yank down my mechanic's costume, the greasy coveralls, ripped t-shirt, Icicle Seafoods baseball cap. After changing, I tie my hair back and look at the face in the mirror—ruddy cheeks, mouth slightly parted, furrow between two nervous eyes.

I start the engine and run it until it's warm. Because used oil is poisonous, I wear rubber gloves. One end of hose goes into a bucket, the other over a tube sticking out of the engine block. I lift and lower the pump handle, and black oil squirts into the bucket.

In band, Mr. K—— rooted out fakers with ruthless zest. Like the other music "hard cores," I practiced oboe during my study halls. Mr. K—— roved from one small practice room to another, checking on us. I'd hear the heavy doors opening and closing. When he pushed mine open, his energy gusted in ahead of his body.

Sometimes, he switched off the light. Sometimes, he stepped behind me and massaged my shoulders, then yanked a chair from against the wall, plopped down, elbows propped on splayed corduroy knees, and asked me to play. I'd been grinding through the difficult sections of the *vivace* movement of Mozart's oboe concerto. "People shouldn't play Mozart until they're older," he said. "To interpret Mozart you need emotional maturity." He swung his ankle onto his knee, tilted back on the chair legs, clasped his hands behind his head. "At least you practice. At least your parents have expectations of you." He lamented the decline of the band program, his single-handedly having to teach every student, conduct three bands, and teach private lessons after school. Once, his best friend had shared the job with him. "Man, those were the days," he said, shaking his head. "We had the best band program in the state. Now, all anyone cares about is sports. How many professional athletes has this school turned out?" We band kids collected money by selling all-purpose cleaner door-to-door. His friend now conducted the best band in the state at a wealthy city school, where his girlfriend was a high school oboe player.

"You know, you're too quiet," Mr. K—— said. "You're too smart. Boys are scared of girls like you."

Later, I wrote in my diary. *Mr. K—— talked to me today about my never revealing myself, keeping everything inside me.* He saw me, things no one else noticed. In my head and in my diary, I turned things he said over and over.

> *Make beak into a knife and swing overhead, turning on right leg to face the west. Circle the platter with right hand. Like a puppet, draw both hands in: hand over hand over knee.*

I babysat Mr. K——'s boys. I never told on them, even when Cole spit in my eye, and Robin threatened me with a pocketknife for not letting him eat a raw hot dog. Afterward, Mr. K—— drove me home. Instead of taking the most direct way, he drove down the wooded road behind the school. "Let's take the long way home," he'd say. Sometimes he'd stop, and we'd sit in the dark. I stared straight ahead, listening to his words. I wanted something to happen, but I didn't know what. Mostly, I wanted him to tell me things, personal things, to make me his confidante.

I was self-conscious around my high school peers, studious, shy, foreign.

Once, I wrote a diary entry confessing love for him. I vowed, then, to perform a kind of maintenance, to keep my thoughts and feelings under tight control. *I'm not going to think back ever again, these pages are part of a closed book, and those feelings are behind a closed, locked, and bolted door in my mind.*

I lift the heavy oil bucket out of the crawl space, and Kathy sets it outside. She hands me a baby diaper, which I shove beneath the engine to catch drips. The filter's tucked under the block, behind the alternator. To remove it, I lie on the floor, my face pressed against the greasy linoleum, and wrap the wrench around the filter and tighten it. When the filter loosens suddenly, my hand bangs against the block. I suck on my bleeding knuckle, then twirl the filter off the shaft, sit up, and drop the filter into the plastic bag Kathy holds out. With my finger, I smear clean oil on the new filter's gasket. To attach it, I must lie across the engine block with my cheek pressed against the heat exchanger.

The engine's warm against my torso. My fingers probe for the threads on the shaft, set the filter against it, and screw it tight. Kathy pours clean oil into the crankcase, and we check the level as it rises up the dipstick. I start the engine, and we shine a flashlight at the filter's base, checking for leaks.

In T'ai Chi, once the postures are memorized, knowledge isn't relevant. The body knows, not the mind. The way, at seventy, Ed the mechanic must have engine schematics imbedded in his muscle tissue, in its cellular memory.

What does the body know?

In ninth grade, my hair fell out, leaving palm-sized bald patches on my head. When it was a third gone, when hair was clogging the vacuum cleaner, my parents drove me to Buffalo, to Chippewa Street. *Hookers shop here*, I thought, staring at rows of wigs, the waved coifs blooming on velvet-skinned, faceless heads. A man slapped a hair-piece over my biggest bald spot. For it to be over, I let my parents buy it. They called it a "fall." It looked like a scalp, like an animal's pelt. I hid it in the dining room closet, and last time I looked, it was still there.

> *Step over log. Open to the north, twisting torso to west. Punch the panda's belly with soft fist, to the west, and then push the panda away.*

On Sundays, Mr. K—— raced his sailboat on the lake. It was a joke among boys in the brass section that he picked only girls to crew for him, and they wore bikinis. In my mind, he took the best musicians, and I wanted to be the best. When he asked me, I convinced my sister to come too. He met us one Sunday, after church. When we climbed into his blue Suburban, the sweet reek of cigar smoke enveloped us. Even now, when I smell it, I turn my head instinctively, looking for him.

My memories of the sailboat race whirl down to a single scene. We're offshore, on the lake. The boat flounders, the sails luffing and snapping in the breeze, the lines askew. We're dead last. Mr. K—— grabs lines from my hands. My sister cries, lying along the narrow side of the sailboat. She's there to make the boat heel over. He bellows instructions. He's never been last in a race, he says.

I see everything from the inside of a Coke bottle. I can't, then won't, move.

Hartalee! I sit, hands in my lap, looking down.

He turns to me, and his words wind up tight, a clockwork coil in my mind.

What's the matter with you? Don't you have a shred of common sense? You better stick to your books, baby, because you will never make it in anything else. You will never make it as a musician.

> *Genuflect to the earth, gently touching right knee to the ground. Read the left palm. Dip tips of right fingers into pool of water. Rise up and strum the harp with right palm. Make a rainbow to the southwest, sweeping arms overhead. Gently squeeze books between palms and push the books away.*

"The universe helps us keep our appointments," a friend of mine says. The first time I opened the plywood cover of *Whale 2*'s engine box, I felt despair at the baffling conglomeration of tubes, bolts, hoses, cables, and loops. I'd never owned a car and had only run outboard motor boats in the past. I studied compression ratios, valve clearances, fuel injection, cooling systems, alternators, but out on the water, the engine shuddering on its bolts, puffing, roaring, all that didn't matter. I didn't know a goddamn thing.

Boat people eventually become fatalists, believing that things happen for reasons. Ravens land on the bow when we're leaving the harbor. A hummingbird or bumblebee circles us once, ten miles offshore, and buzzes away. My wedding ring falls into the bilge. A killer whale swims beneath us with a seal in its jaws. Despite routine maintenance, the boat breaks down. We ask ourselves, *What does it mean?*

If the sailboat race is an appointment I kept, then so is this circling back to a boat, to tools, to animals I can't learn about from books, only from watching, to a cantankerous engine, to a wound-up voice in my head, forewarning me of disaster.

As I pull the rusty grease gun from its bag, I tell Kathy about the wrecked boat my friend found it on, how that friend taught me the particular posture necessary to apply grease to the zerk fittings for the steering shaft journals and bell housing. Lying across the engine block, a flashlight in my mouth, I place the grease gun behind the battery case. With my right hand, I probe through globs of old grease for the metal nipple, maneuvering the tube up behind the turbocharger. Bending the hose, I work the metal end onto the fitting until it clicks into place, then pump the gun handle three times.

Sink back on left leg, facing east. Sweep right palm past right ear and to the east. Sweep right hand, palm down, to right hip. Repulse the monkey back through the jungle, pushing the palm tree leaves away.

When Kathy practices T'ai Chi, her movements flow the way the thousand movements that make up an eagle's lift-off from its perch and its flight appear as one continuous motion. I'm like the immature eagle I saw once, hanging by one leg upside-down in a tree, eyes bright, beak opening and closing, flapping its heavy wings.

But in T'ai Chi, it's never performance. It's always practice.

I practiced in the bathroom. Scales. Arpeggios. The *Vade Mecum of the Oboist.* I tied blue, green, or red nylon thread to table legs and wrapped oboe reed cane to brass tubes, then whittled reeds while watching TV. Maintenance, for the oboist, requires almost constant reed-making, an art unto itself, to replace reeds as they wear out. It's constant because one in ten reeds sounds good, and each one lasts two weeks, at most. Once, when I cracked a favorite reed on my tooth right before an audition, I knelt by my bed. I prayed for God to repair the crack. I opened my eyes very slowly, and, before they focused, believed for a moment that the reed was whole. I still have a scar on the middle finger of my right hand, a little half-moon, from a swipe of my oboe reed knife.

Summers, I traveled to music camps to study with members of professional orchestras. When I ranked fourth out of five oboists at the Saratoga fine arts camp, I vowed to practice three hours a day. Mr. K—— told me that for every day I missed, I lost three days of ground. I calculated the losses in my mind. I was swimming backwards. But what I remember most that summer was watching a dozen girls, other music students, run outside screeching and laughing during a thunderstorm to shampoo their hair in the rain. *I don't know how much longer I can take this loneliness,* I wrote in my diary.

When I came home at the end of the summer, my sister baked me a cake decorated with the words: *To my sister the oboe player.*

Grasp the stick on left hip. Part the wild horse's mane, moving side to side. Repeat.

While Kathy, rubber gloved, bug-eyed in safety glasses, checks the battery electrolyte, I crawl into the space beside the engine and shut off the fuel flow. Using a hammer and screwdriver, I loosen the housing lid and pull out the old paper filter, which is black with dirt. To clean the filter bowl, I drain the fuel into a tin can, remove the bowl from the housing, and swipe it out with paper towel. I screw the bowl back into place, insert a new filter, pour clean diesel to top it off. As I ask Kathy about the battery charge, I keep pouring and pouring diesel and the level never reaches the top. Then I realize that it's running out the seam between the bowl and housing, where it's not seated right. I grab a sorbent pad to soak up the diesel that's spreading under my boots and dripping into the bilge until my hands glisten and I'm nauseous from the smell. *Damn it,* I hiss. There's nothing to do except start all over again, clean up the mess.

"Breathe, Eva," Kathy says, handing me tools. Slowly, I retrace my steps, then replace a second filter, clean the lift pump screen, raise and lower a lever to clear all the air from the fuel system.

Each step in maintenance requires concentration so that gaskets are in place and lubricated, so that unfiltered fuel's not passed into the injectors. An entire chapter of my diesel mechanic's manual is entitled "Cleanliness Is Next to Godliness." Spilled fuel ends up in the bilge, and if not cleaned up, eventually spews into the ocean. It's all connected.

"Done," I say, as I let the plywood cover drop back down over the engine. For effect, I brush my hands together, and then we stash our tools away for another hundred hours.

Just once, I want to see you relax for a few minutes. Just once, I want you to stop judging yourself, Kathy said as I lay on my back on an exercise mat after T'ai Chi class. My eyes were closed, and she sat on her heels beside me and pressed my shoulders toward the floor.

> *Lead with left toe, hiding right fist with left palm, three-quarter turn around the world. Swing right leg. Extend right heel to sky, parting the curtain.*

The first time I drank wine, at a party the summer before my senior year of high school, I swayed back and forth on a wire in a grape vineyard. Other kids laughed in surprise, called me "Loady." Later, I walked down a dark farm road with a boy I'd loved secretly for a year. As we sat on the thruway overpass watching trucks thunder by, he told me that he hated his life. Once, his mother chased him around their trailer with a knife, once with an overturned chair. His father had been having an affair. If I told him anything about my family life, I don't remember it. We made a pact, that if we ever felt suicidal, we'd find each other. At home that night, I couldn't sleep, so I wrote in my diary, then ran outside. A warm wind swept through the maple leaves, crickets and tree frogs roared, stars throbbed dully in the humidity. The air smelled rank, wet with possibility.

Like the whales, the engine's cagey. Despite routine maintenance, the unexpected happens. As Kathy and I guide the boat between two islands, I think I hear a high-pitched whine, like a whistling kettle. Suddenly, an alarm jangles. When I yank open the cabin door, the ringing fills my skull, and a cloying smell of burnt antifreeze wafts from the engine.

"Shit!" I shut it down and yank out the connection on the alarm.

"What's going on?" Kathy asks.

"The engine's overheated. The saltwater intake must be clogged."

We look around. The boat drifts toward a rocky island shore, but it's too deep here to drop the anchor. I remember watching a friend do this once. After digging frantically through the fish hold for the dry suit, Kathy helps me pull it on over my clothes. Feet-first, I slide off the boat's side into the water. Near the stern, I duck my head under. It's so cold my jaw bones ache as I grope for the metal screen covering the intake, scraping off eelgrass and a mat of algae. When I resurface, Kathy grabs my hand and heaves me aboard.

Inside the cabin, we read the troubleshooting chart aloud. We pick algae and tiny shrimp from the saltwater strainer bowl, then unscrew the pump cover to check the impeller, and water pours out—it didn't run dry, and the impeller's intact. A bad overheat could easily crack the head on this engine, I tell Kathy. While I reassemble the pump, Kathy goes outside to see where we are. "We're getting pretty close to shore," she calls. "Should I drop the anchor?"

"Let's just see if it starts," I say. My fingers on the key, I breathe, turn it, watch the needle on the temperature gauge creep down.

As we angle the boat into deeper water, my legs tremble. All day, I picture us going up on the rocks, finding the impeller destroyed, pieces of it sucked into the engine, pistons frozen into place, the engine seized up and useless. Though it feels neurotic, I've heard this is how the mind releases us from traumas, replaying them over and over until their energy's gone.

Genuflect to the earth, softly touching left knee to the ground. Dip the tips of left fingers into pool of water. Rise up and strum the harp.

During my freshman year at Northwestern University, I studied with Ray Still, who played principal oboe in the Chicago Symphony. Every Saturday, I rode the elevated train to his house for my lesson. I never knew what to expect from him. He was in his sixties, balding but trim, with intense silver eyes that burned though my skin and bones. At my first lesson, he pointed to a chair for me to sit on. I dropped my reed into a film canister of water to soak and, with damp, shaky palms, assembled my oboe. As Mr. Still clattered around the kitchen, he called, "Let's hear your warm-up." I shook the water from my reed, stuck it in the oboe, then ran rapidly through scales and arpeggios. He strode in, grabbed the oboe, placed the reed on his lip, and blew a single note that intensified until my ears rang, until the room closed around me like a vibrating organ pipe. "Just play one note," he said. "Hold it. Feel it push up from your gut and intensify until it fills your head. Until you don't hear just that note, but harmonic tones ringing in your ears." Then he played a recording of a nagasfaram, the oboe's Indian ancestor, and told me to imitate it. When I hesitated, he ordered, impatiently, "Come on, do it, just wail."

That's how he taught me to sing through the oboe, to imitate Kathleen Battle, to listen to the songs of Schubert, to mimic a Chinese flute by blurring the distinction between notes. "How do you walk up stairs?" he asked me once. Baffled, I replied, "I usually run up them, two at a time."

"No. You take each one deliberately, breathing from your diaphragm. Everything you do is for the oboe." He showed me how to rub my index finger down the crease between my nose and cheek, how my oiled fingertip then slid from one key to another.

One afternoon in spring, we'd been working on Schumann's songs. "I want you to play all three songs, one after the other. I'll tell you when to stop." I moistened my reed in my mouth, looked down at it, squeezed its opening wider. I'd never played even one song through without stopping for air. Mr. Still adjusted knobs on a reel-to-reel tape recorder perched on a stand. As I played, the tape turning silently, Mr. Still paced around me, his head bent. He swept his arms. The waves of his moving arms lifted sound, like a sunken ship, from my body. The source of my breath seemed limitless. When I finished the third song, before I had a chance to stop, he shouted, "Again."

I played the three songs through a second and a third time. Suddenly, he shouted, "Stop. Listen to this." He strode over to the recorder, rewound it, played it back. I didn't recognize what I heard. "That's you," he said. On his brow, a mist of sweat glistened. "That's how you can play."

Goddamn it. I close my eyes and cover them with my hands. It hurts behind my eyes, in my shoulders. I get up from the bunk and stare out the dark cabin window, listen to the chimney cap spin in a wind gust.

"What's wrong, Eva?" Kathy asks.

We've been holed up all day in a storm, reading, writing, talking. I've been telling her about Ray Still, how his students called him a magician because he could pull music out of us, music we could never recreate in our practice rooms, how he told me I'd improved more than anyone else that year.

"I was a musician, Kathy. Why did I quit?"

"What happened?"

I described how one night, as I walked back to my dorm from the music building, where I'd practiced all afternoon in a tiny room, I stopped on a wooden walkway over a stream. I could still hear a chaos of operatic voices, trombone riffs, violas, cellos, drums behind me. In the humid spring air, platters of light wavered and drifted on the water's surface. In that spot, I felt no self-consciousness or isolation, like I could belong somewhere. As I pressed my oboe case against my chest, a voice in my head I'd never heard before said, *You don't have to do this. You can just stop.*

One evening, we watch killer whales feed. The adults dive deep to drive salmon to the surface. They chase the fish in tight circles, sometimes pinning them right against the boat's hull, and calves and juveniles chase each other. Kathy's listening through headphones, concentrating on subtle adjustments to her digital recorder. She looks up at me and gives me a thumbs-up, mouths, "This is great," then hands me the headphones. My head fills with a cacophony of shrieks, whistles, and so many echolocation clicks that I imagine a mad underwater typing pool below us, whales in nun's habits battering the keys. No engine noise from other boats. No current dragging the hydrophone. No waves churning. Perfect. But the sun's down, and we have to find a good anchorage. There's a small craft advisory for tonight. After the whales pass and the calls die down, I suggest, "Let's move on ahead of them and record one more time," and she switches off the tape deck and pulls in the hydrophone.

I lean inside the cabin to start up, but when I turn the key, nothing happens.

I push up the engine cover and shine a flashlight over the batteries and check the connections, check the fuse box. Nothing looks wrong, of course. Most problems on diesels are electrical. Maybe it's the starter. I replaced it last spring.

"Maybe we should call for help. It's getting dark," Kathy says.

"I don't think there's anyone out here, Kath." We're sixty miles from town, twenty-five from the fish hatchery.

Kathy digs our spare starter out of the fish hold. It weighs thirty pounds, and, as fisherman say, will be a bitch to deal with when I'm scrunched into the cramped engine compartment. As I crawl in, hefting the spare starter, I remember another mariner's philosophy, "Try the simplest thing first."

I perch the spare starter on the table and tighten the bolts holding the old starter to the block, just in case they've worked loose. "Try it now," I tell Kathy, and when the engine shudders and roars to life, we hug each other and cheer.

Move toward ball, take it, then bring it back to the river and squat low on left knee, right leg extended, pointed toe. Carry ball back up, part the river, and push the air away.

I didn't recognize that voice, just that it was true. I heard it again the night I watched a film about wolves in Alaska and knew I needed to come here. I heard it the first time I saw Prince William Sound from the scratched window of a floatplane, and it told me I was home. I heard it when I saw my first killer whale, when I left my marriage, and I listened to it six times one year, each time that it told me to fly back east to see my mother after her brain aneurysm. At nineteen, I followed it away from music school, back to the state college near my hometown where I studied forestry. I never told Ray Still I wasn't coming back.

Like evolution, the voice didn't always lead to higher states. But it was my voice. And maybe Mr. K—— was right, maybe oboe players go crazy. For three years after I put down my oboe, I drank, had an abortion, tried every drug offered to me, woke up some mornings horrified at who lay beside me in the bed. I nearly failed organic chemistry, physics, and math analysis, but still managed to graduate college with honors. With a boyfriend, in a tiny car, I hauled my oboe up the Al-Can highway to Alaska. Once in a while, I'd pull it out and play when I was sure no one could hear me, but it was often too painful, and I'd stop.

For ten years, I traded wardrobes, from homemade dresses to hippie clothes, to coveralls and flannel shirts. Oboe player, Deadhead, biologist, writer. But music isn't imitative, and it isn't a style. It isn't even a choice. It's inside you like a root, like a cluster of cells.

The next morning, after fixing the starter, Kathy and I cross Montague Strait in worsening weather. We head for a calmer passage to wait out the storm. Just to make sure we're not missing whales, we drop the hydrophone to listen. I put the boat in neutral, lean into the cabin, and turn off the key. The engine keeps running. "What the hell?" I murmur, staring at the switch panel.

"What's wrong now?"

"You won't believe this," I say, "But the engine won't shut down."

"What does that mean?"

"I don't know; it's never happened before." I remember a phrase a mechanic told me once: *If something goes wrong, think of the last thing you did to the engine, the last thing you touched.* I remember the fuse box.

I rush to the front cover of the engine and pull it off, releasing the full force of the engine's roar into the small cabin. When I unscrew the lid over the fuses and pull out the main one, it falls to pieces in my hand. A tiny spring flies onto the floor.

"Here's the problem," I shout to Kathy. My hands shake, and I drop more pieces of the fuse.

"Slow down," says Kathy, with alarm, grabbing a spring before it rolls into the bilge. The boat rocks heavily back and forth in the waves.

But I don't want to slow down. I grab the spring, and after several attempts, I put the fuse back together and bind it with electrical tape. Kathy turns the key, and the engine shuts down. I sit on the floor, laughing and crying.

Form gentle fist with right hand. Punch the panda's belly and softly push the panda away. Hiding the fist with the palm, turn one-half way around the world.

That night, sipping tea in the bunk, Kathy says, "When I'm teaching the form, and people are watching me, it's hard not to be nervous. I'll start a difficult move, and my mind will slip, thinking, 'what if I can't do this,' and my leg will tremble. And I'll let it go, and the move is fine. I'm always forgiving myself."

That's the hardest part. Forgiving yourself. Not just for the big things, but for the smallest.

In my late twenties, during a visit home from Alaska, I pulled out my cardboard box of letters, yearbooks, and high school and college journals and sat up until 3:00 a.m., forcing myself to read. I cringed at my teenaged prayers to Jesus, at my vows to do good deeds without telling anyone, at my rigid to-do lists, my unsent letters and sappy poems to secret loves, my confession of love to Mr. K——, my incessant whining about loneliness. I recoiled just as hard at my college journals, chronicling my slippage from perfectionist to out-of-control party girl, and I stuffed the whole pile into the kitchen garbage can. I was a killer whale biologist now, a skier, winter mountaineer, vegetarian. I didn't drink. I'd recently begun playing oboe again, in a trio with two recorder players. Mostly, I wanted the past to erase me—this me—out of its narrative.

A few years later, driving along Lake Michigan to visit a relative, I passed unexpectedly through the Northwestern University campus. I slowed down, staring at the lake and walking paths, then pulled over and parked. Navigating toward a dissonance of cello, oboe, voice, trumpet, of scales, trills, runs, arpeggios, I stopped outside the sharp-angled building with its slit-like windows. A trumpet player sat on a stone, improvising. It seemed, if I walked in, I'd see people I knew once, chatting in a lounge, their instruments in their laps, the bassoon player, a red-faced, big, blonde guy, and the hottest oboist, who made Ray Still's reeds, telling a story of jumping a train, riding it into Chicago on weekends. When I turned and walked the cement path along the lake, cold wind snapped through my hair, and whitecaps streaked the water, indigo and sharp as stained glass.

I knew where I was going. Ahead of me, a small bridge arched across a stream, and I imagined myself, at nineteen, standing there, looking down, a girl who twiddled away on technical exercises in the practice rooms, who prayed for cracked reeds to mend themselves, who once sliced a half-moon off her knuckle with her oboe reed knife. I imagined taking her arm, walking her back to the car with me.

Retreat the heel. Great high palm on right brushes palm leaves away from face. Great high palm on left brushes palm leaves away from face. Open.

Several years ago, in a phone conversation, I confessed to my mother how much I regretted throwing out my childhood diaries.

"I have to tell you something. You have to promise me that you won't be mad at Dad. He found them in the garbage can. He took them out and put them in the cellar somewhere."

The next time I visited, I found them in my father's basement library, wedged between dictionaries in one of the glassed-in shelves. While everyone slept, I sat on the high bed in my room and read.

> *I wish that he would hate me. I want him to respect me, but I wish he would not like me, personally. I'm even afraid to feel anything because he might know how I feel. I think he does. I can't talk to anyone about this.*

During my senior year of high school, our rag-tag high school band ground through Shostakovich, Rossini, Ravel. We performed the 1812 Overture, complete with a simulated cannon, someone shooting blanks into a wine barrel. During concerts, rivulets of sweat poured down Mr. K——'s face as he conducted, gesturing wildly with his baton and body. I was afraid he'd fall backwards off his podium. Oval stains spread on his shirt, as if by sheer exertion he could create the band he wanted us to be.

Once, I saw him do just that. I was playing in the county's summer orchestra, led by a gaunt old man who conducted with a gentle lifting motion of the baton between thin fingers, like someone trying to paddle a canoe with a teaspoon. His voice was tired, half-lament, half-sigh. And, though we were top musicians in our schools, the orchestra luffed. When he stopped us, for seconds we didn't notice, and instruments kept playing, dropping out one by one, like a Victrola record dying down.

One day, he was sick, and Mr. K—— stepped in to substitute. He raised the baton, unleashed us with a push of both arms, rocked his torso back and forth, his head turned sideways, sweat popping out on his forehead, his arms pumping, cuing the percussion with an intense look and arm-swipe, his cotton shirt blotched with sweat, and it was like someone doused our limp bodies with a cold hose until our skin tingled and twanged. Suddenly, we were real musicians, and he was the magic man. His arms wound us up like alarm clocks until we chimed. He loved it, I could tell, because he beamed at us, never screamed or cursed or heaved his baton. Remembering this, I now see his daily frustration with a limping band program, apathetic students, all the rest. Mr. K—— was a *musician.* I see his passion all twisted inside, how, in the sailboat race, his rage wasn't about us at all, but about his own thwarted desires.

The last time I saw him, I was visiting my family. One day, I asked my mother about an oboe teacher of mine, Dr. M——, a professor at the nearby university. I'd studied with him in preparation for my college auditions. He'd taught me the rigorous repertoire of the serious student: the Barrett oboe method, the *Vade Mecum*, difficult passages from Bach, orchestral studies, sight reading, and singing. A serious man in wire-rimmed glasses and formal clothes, Dr. M—— had never asked me a personal question. I'd never learned anything about his personal life.

"We've always been afraid to tell you."

"Tell me what?"

"It happened right after your friend killed herself. We were afraid you were too sensitive then, that it would upset you." My mother stood in the living room with the dust rag wrapped around her hand.

"Mom, that was five years ago. What happened to him?"

"He became not right in his head. It was some kind of depression. He shot his mother and then himself."

I stared at her. Her words floated like ice pans. She bent to dust an end table.

"Why didn't anyone tell me?"

"It's over. It happened five years ago."

I didn't know where I was going when I pushed open the screen door and headed down the driveway. As I turned right onto the road, starlings rasped on the telephone line. I walked to the high school. When my hand touched the thick, metal door knob, I almost turned around.

Mr. K——'s face lifted in surprise when I walked in.

"It's my favorite student," he exclaimed. "You look great. Look at me; I'm an old man." His once-blonde hair was silver but still draped boyishly across his forehead, and he pushed it back with a familiar sweep of hand. "Come on in and sit down. How the hell are you?"

I could hear a clarinetist running through scales, a trumpet blaring from the practice rooms. In a semicircle, around the wooden conductor's podium, dented metal stands tilted like ravens in front of blue and tan chairs. "I came because I just heard about Dr. M—— No one told me until today. I needed to talk to someone who knew him."

"You didn't know?" He slumped back in his chair. "I'm still sad about that. It was a tragedy. A bunch of us played at his memorial service."

"Why? Why did he kill himself?"

"No one knows. I guess he just snapped. His mother had Alzheimer's."

We sat still for a moment.

"Do you still play your oboe?"

I hesitated. "Sometimes."

He talked about the band program, which was worse than ever. Soon he'd retire and move away. "You were one of the best. Students like you and your sister made it all worth it."

When he hugged me good-bye, I smelled the sweet, tar smell of cigar smoke. He gave me the same line as always. "You're getting too thin. If you're not careful, there won't be enough left of you to love."

Nothing is what it once seemed. T'ai Chi's helped me to reconcile, or simply approach, some of the paradoxes in life, because T'ai Chi itself abides by a paradox. When you practice the form, you think nothing, not of future or past. Even so, the quality of this present practice is made up of all past practices, of everything in your life.

> *Draw left side in to face the east. Pat the horse's flanks with palms. Genuflect to the earth, sweeping arms wide to circle the eastern sky.*

My oboe is still my itinerant companion. And now I have an oboe student, a talented, quirky girl in junior high who loves to play duets with me. Sometimes I play in the regional orchestra. One summer, I took my oboe on a rafting trip on the Copper River and that fall, performed in three concerts with my friend David, accompanying his voice and guitar. One of our favorite duets is an Irish air called "The Song of the Books." David's the one who said the universe helps us keep our appointments, and obviously he's right. Still, sometimes, I want to strike my forehead against a wall when I imagine a life of music. Instead, I bring my oboe on the boat. Once, at night, after my colleague Craig and I had searched unsuccessfully all of a short winter day for the AJ killer whale pod, which had been seen every other day of the previous week by some boat or other, we heard them, in darkness, on our hydrophone after we'd anchored up. "Of course, here they are now," Craig lamented, and their chattering, indeed, seemed cunning, mischievous. I pulled out my oboe and tried to imitate their calls. I'd listen with my eyes closed, then hum the phrase, but I couldn't find all the notes, so I played Mozart for them, and "The Song of the Books," and the next day, as they milled around our boat, I sat on the bow playing while Craig recorded their sounds, and one whale, for a moment, seemed to come closer to listen.

Straddle stream. Squat low under tree limb as you cross hands, palms facing sky. Sweep arms out to sides. Face north, palms facing each other. Balance the leaves on the sides of your hands.

The act of listening is a key to maintenance. One chapter of my diesel mechanic's book is devoted to "Smoke, Knocks, and Thunks." A boat person's ears become finely tuned to the particular rhythmic tapping of pistons, to the clunk of gears shifting, to the vibration of the lower unit with the wheel hard over.

A therapist once told me that she didn't think I'd ever really heard the sound of my own voice. She told me I needed to go deep into the woods and scream. I did that, but I sounded like a throttled owl, and it just scared me. So now, instead, I pull out Schumann's three songs, the pages soft and torn at the corners, Mr. K——'s slanted handwriting slashed across the top: *sound—musicianship—vibrato*. I put the oboe reed on my bottom lip, form the gently muscular oval—the oboist's embouchure—around the reed, and blow a single note.

Sound begins as a suggestion, an idea, a shift, like the beginning of wind. The body breathes in a lake of breath and holds it. It rises back up in a stream, expands and fills the throat, then constricts again, under great pressure, through the tiny opening between the twin blades of the oboe reed. The sound emanates from the facial bones. The skull opening like a whale's throat, harmonic tones thin as acupuncture needles pierce the eardrums.

Straight leg becomes a rising sun. Extend leg as hands cross in front of face, palms facing toward you. Turn palms toward east and sweep arms wide to circle the eastern sky as you squat under the clouds. Body and left leg rise like the sun.

It's our last day on the boat for the summer. A storm's just passed through, dropping thick fog across the Sound, so the water's milky and still. Kathy and I drift off Point Helen, our hydrophone down. For the past half hour, we've been listening to killer whale calls growing gradually louder. We can't go looking for the whales, or we'll be lost in the fog, so we float in sight of land and hope the whales come by.

When the sounds are loud, Kathy hooks her tape recorder to the hydrophone. I've never heard these particular calls before, these wild, banshee cries, reverberating bell-like, *whoop-whoop-whoops*, these foghorn-like blasts and piercing screams. They remind me of what, according to the Tlingits, seals say when killer whales arrive: *Here they come. Here come the warriors.*

Suddenly we're surrounded by the whoosh of breathing. We whisper. Should we follow them, try to get photos? I recognize a whale called Jack, his dorsal fin curled over like a question mark. We're reluctant to move, to break this magic. These calls are so strange, not their normal calls. Maybe they've come to say good-bye. In the end, we just let them pass, let the fog drape its shroud around them.

"Eva," Kathy says, still whispering. "Don't you see? Here you are, studying whale sounds. This is music. You've come back to music through science."

This is what I see now, and it, too, is a kind of maintenance and a kind of paradox. I want to live in the present moment, but I swim through thick kelp beds dragging flotsam behind me. To cut away the old forces holding me, I have to dive below with a knife. Before I slice those kelp ribbons, rotten crab lines, old nets, I have to see them clearly for what they are.

Lift up the waterfall. Let your palms flow down the waterfall together in peace.

Leaving Resurrection

He who is to see what he has seen before—his eye quivers.
He who is to hear what he has heard before—his ear rings.
He who is to have sorrow—his throat feels full.

—Aleut sayings

Fifty miles offshore in the Gulf of Alaska, in water a thousand feet deep, a man fished alone. Fishermen call that underwater canyon at the continental shelf's edge "the trench." Along the trench bottom lurk groundfish—halibut, cod, skates.

The water surface rollicked. Harold's longline, spiked with hooked fish, slowly lifted. He'd listened to the weather forecast in the morning—southeast winds to twenty-five knots—but he estimated it was blowing thirty-five, a gale, by the time he brought in the last fish. When he worked his line, Harold couldn't hear the radio above the wind. If the Coast Guard had issued a weather warning that afternoon—a change in conditions—he didn't know it. Later he learned it was blowing fifty.

He'd put out stabilizers—aluminum fins hung port and starboard off wooden booms—to ease the boat's pitch. The *Rocinante* was thirty-three feet long, each wave nearly twenty feet tall. When he turned toward Seward, the hold was stuffed with halibut and cod. "I just misjudged how hard the wind was blowing," he told me, months later. "In six more hours, I would have been in Resurrection Bay."

A wave drove her on her side, shifting the load of fish. With the stabilizers down, he couldn't right her. He'd done that once, off the Oregon coast, bringing *Rocinante* back from Hawaii alone in a storm. He'd turned the wheel hard-over, and somehow, miraculously, she'd popped back up. But this time, luck wasn't with her. A boat fishing nearby heard his Mayday and picked him up an hour later. *Rocinante* hadn't sunk yet. Harold clung to her wooden hull like a barnacle. I imagine his rescuers having to pry him off.

For months after I hear the news, I imagine her. I dive down fathoms of water to the trench bottom. I shine a flashlight over the curve of windows. Fish glide through the smashed glass, mud plumes billowing as they brush her sides. Enormous pressure pins her down.

Harold fished mainly in the Gulf, but during storms, he'd come into sheltered water. On days when wind shoved squalls down the passages, I'd look for her. Passing cloudbanks swallowed islands, snagged in trees. After they passed, I'd suddenly see *Rocinante* swaying in front of camp like a ghost ship, as if she'd drifted in unmanned. It was her look, her wooden hull with its rounded stern, her pilothouse perched like a cupola on the bow.

When he retired from his town job, and after his wife left him, Harold drew up plans for *Rocinante*. He styled her after the wooden fishing boats his family designed and manufactured on the Norwegian island where he was born. Harold built her himself.

To me, she was fanciful, with her graceful lines, wooden pulleys and wheel, the curved pilothouse. But for him, she was practical—a work boat—with an electric galley, hydraulic systems, stores of canned food. What wasn't practical about her was her size. Though seaworthy, she was small for a boat that fished all winter in the Gulf of Alaska. Harold was always on the lookout for a bigger boat so he could fish in rougher seas. When the time came, he would have sold *Rocinante*. In her cabin hung a brass plaque with a quote from *Don Quixote*: *The best carriage beast in the world.*

Harold's family sold their land and business after two of his brothers drowned while fishing. It's now a museum. But, like an island, the sea defined him. For decades he worked as an accountant for a seafood distributor in Anchorage. I know nothing about his marriage except that its end prompted him to launch the scheme of the *Rocinante,* late in life. And now she's gone, too, and Harold has a new boat, also named *Rocinante*, an entirely utilitarian fiberglass boat—nothing quixotic about her—at least not yet—large enough to weather storms the wood boat couldn't. Such is luck, or its lack. For some, fate is entwined with the ocean, and for Harold and for me that seems right.

A week before Easter, we're anchored up several miles out of Seward, in Thumb Cove, a tuck in Resurrection Bay. Craig's seine boat, the *Lucky Star*, slides up and down, silky on the faintest aftertaste of swells. Sitting on the flying bridge, I write in my diary. Rain blows in from beneath the roof—the "sissy lid," fishermen call it—and peppers the pages. Where a drop lands on my writing, a word floats. I look up and watch gusts pucker the steel green water. Passing under the boat on their way to the beach, swells unfurl like white scarves.

Below, in the cabin, Craig and Lance prepare supper. We're settled into this anchorage for the night, a storm in the Gulf having thwarted our making the fifty-mile crossing east through open water to Prince William Sound. We're geared up for ten days of research, hoping that killer whales will follow the spawning herring and their predators—seals and sea lions—into the Sound. Because of the arrival of herring, a bulwark of the Sound's food chain, April is a good month for whale research, but we have scant information on killer whale feeding habits during this time of year because of the weather.

This April, winter isn't ready to lie down yet. I'm relieved to be anchored up here. Over the years I've worked on this project, Craig's boat has become as familiar as my cabin in Fairbanks. Though we have other lives in town—families, houses—when we work together, we become a kind of family. Home is this forty-two-foot space.

On shore, varied thrushes call, a strange buzzing like high-tension power lines. That sound signals spring in south-coastal Alaska, but in the Gulf, just outside the entrance to Resurrection Bay, big storms still pass, one after another, born hundreds of miles to the west, in a weather gyre that spins all winter over the Bering Sea. One of those storms drove us back here tonight.

We left Seward earlier this afternoon, after a full day's preparation, hoping the weather forecast was wrong, the storm delayed or weakened before it reached us. The forecast had called for fifty-knot winds. Lance checked his fancy watch that displays barometric pressure. The previous night, it had read 1018 millibars. By midday today, it had dropped to 1006. With such a significant change in pressure, we knew a storm was coming, but not when. We decided to go for it.

Our choice seemed right as we headed toward Cape Resurrection, not much wind, not much swell coming in from the Gulf.

"If we don't make a break for it before the storm hits, we may be stuck in Resurrection Bay for days," Craig lamented. We sat outside on the flying bridge, Craig at the wheel, Lance and I sitting on either side of him. Rain squalls came and went. A group of Dall's porpoises bow-rode the boat briefly. *A sign of luck*, I thought.

I matched the cloud-draped landmarks we passed to the names on the nautical chart of Resurrection Bay spread across my lap: Thumb Cove, a headland named the Iron Door, Humpy Cove, steep mountainsides with Prospect, Spoon, and Porcupine glaciers hanging down, and a slim passage called Eldorado Narrows. Past the narrows, around Cape Resurrection, the bay opened out to the Gulf of Alaska.

As we neared the cape—thousand-foot cliffs jutting up at the Gulf's edge—a smudge darkened the water at the horizon. Looking through binoculars, I made out foam streaks, wind raking the riptides off Barwell Island. When we entered the wind line, we went below into the cabin. Craig sat at the helm, peering at the radar, and Lance read a novel.

The *Lucky Star* began to lurch. Heavy waves washed over the bow, sending up curtains of spray that streamed down the cabin windows. Feeling nauseous, I crawled into a bunk. The boat heaved, and my hand flew up to the ceiling to hold me in place. Suddenly afraid, I jumped back out of the bunk to see what was happening.

"Are we going to keep going?" I asked.

Craig is normally very cautious about crossing the Gulf. Lance, who seemed unperturbed, is more adventuresome, has piloted his own boat in the open ocean off Vancouver Island.

"Probably not, but I just want to get to the cape to see what it's like," Craig said. "I knew we should have left earlier."

The *Lucky Star* has a tunnel hull, built for shallow water where salmon congregate. That hull shape—a deep V at the bow and nearly flat beneath the stern—makes her a poor sea boat, and she slides and rolls in ocean swells. Craig doesn't like to be out in seas over twelve feet if the wind's blowing. In calmer conditions, the *Lucky Star* bucks a head sea pretty well. Although waves break into white crystals against the hull, she powers through them.

At the cape, the swells were twelve feet tall, overlain with wind chop. The sea surface was a huge lung, rising, falling. Along the coast, swells are the echoes of distant storms, spreading thousands of miles from their origin. Storm swells intensify like deeper and deeper breaths.

Wind-chop pants, like the breath of panic, like my breaths as I looked from Lance to Craig. The *Lucky Star*'s bow plowed into each wave, and she wallowed for a moment before regaining her heading. With one hand I held onto the table, with the other, to the back of Craig's seat. Lance put down his book. Between large waves, I staggered around the cabin, trying to stash away anything that might fly off a counter or bunk. Tea cups slid back and forth across the table. A pair of binoculars crashed to the floor.

Having a purpose helped. After clearing away the loose gear, I climbed up into one of the seats and watched the sea out the window. I imagined what we'd look like from a distance, the bow plunged beneath the water completely, the cabin lights cut, my face blotted out.

The swells moved with a brown bear's gait. They smacked against the cliffs, falling back, confusing the seas into gray-green heaps. Craig swung the *Lucky Star* around just off the cape.

Going with the seas, our movement smoothed out. Without adrenaline to distract me, I felt seasick again, so I grabbed my diary, pulled on a jacket and went outside. Lance watched me through a porthole. I moved carefully, holding onto the boom and rigging as I climbed to the flying-bridge. A freak wave might have knocked me into the sea. Breathing cold air, I felt nausea leave my body. Knowing we weren't going out in the gale, I acknowledged the beauty of the burly ocean, the wind tearing spume off wave tops. It was all so much more immense than our plans.

In a storm, the ocean enlarges beyond any science. Alaskan weather forecasts are notoriously inaccurate. Larger ships receive daily "weather faxes," showing maps of storm centers, high and low pressure systems. Seafarers learn to interpret cloud signs, a barometer needle's slide, aches in joints, movements of birds. They touch wood, for luck.

When I look up from writing, the sky has gone grayer. I think about our research project, why I like it so much. As a writer, gathering data is also my aim. I'm patient, and besides, it takes decades to learn about the whales. Half the time, they're not there, or the weather's bad. To do this work for long, you have to love the sea and its temperament as much as the research. You have to love just being out here.

In the traditional beliefs of the Nootka people of Vancouver Island, a horrifying monster lives beneath the sea. Sisiutl seeks those who can't control their fear. *When you see Sisiutl, you must stand and face him. Face the horror. Face the fear.*

As I watch squalls move through our anchorage, eagles lifting and landing at the stream mouth, I think of the hour Harold spent hanging onto the capsized *Rocinante*, knowing it could sink at any moment. I think of my own fear out at the cape and of Lance's confidence. What lives in the place between facing one's fear and being arrogant? What lives between cowardice and caution? There's experience—something made of more than knowledge and hours—and then there's luck.

Anthropologist and writer Richard Nelson says that luck is essential for a Koyukon hunter's success, regardless of skill. Arrogance or oversight can cause people to lose their luck. The Yup'iq people of western Alaska believe that luck comes from sharing, from being looked upon favorably by the earth and animals because of correct behavior. Are people who follow this worldview fearful? Or, when luck is lost, do they accept consequences, try to figure out what went wrong?

"Dinner's ready!" Craig calls up from below. As I close my diary, I think of the Gulf of Alaska, around the cape. The gap between my experience and my knowledge is that wide.

The next morning, the weatherman's laconic, nasal voice drones the same forecast as the night before, adding grandfatherly emphasis when he cautions, "Storm warning, repeat, storm warning. Northeast winds to fifty knots, seas to twenty-two feet, rain and snow."

Through my binoculars, I watch spume rip across the surface of Resurrection Bay, outside our anchorage. No one suggests leaving. Laptops humming, Lance and Craig sit across from each other at the table, their hair sleep-disheveled. I tease them for wanting to stay inside, a couple of nerds.

After pulling on warm clothes and raingear, I kayak to shore. Although there's only a slight chop in this protected cove, to make progress I have to stroke with all my might against williwaws that stampede down the steep slopes. Rain pellets sting my face raw. After carrying the kayak up the beach, I tie it to an overhanging branch above tide line. I look back. Around the swinging *Lucky Star*, gusts land on the water like punches.

I walk north along the shore. A quarter mile down, I spot something sticking up from last year's dead grass. As I get closer, I see it's the top of a wooden boat's wheelhouse, high above the tide line, surrounded by spruce saplings not much taller than the empty window frames. When I reach the wreck, I kneel down next to it. The curved wheelhouse is tiny, probably from a skiff. A single person could have stood in it, peering out at the sea while steering. The lower edge of the wheelhouse has sunk into gravel. Curls of green paint cling to the weathered planks. On top, at a slight angle, a stovepipe fitting remains, the pipe gone.

I imagine the whole boat buried in the gravel below the wheelhouse. I walk behind it, then crawl inside. A narrow ledge below the window frames is filled with pebbles. Once, a compass might have swayed there on its binnacle. Saplings screen the view, their branches poking in through the windows.

Mariners say that live plants on a boat are bad luck—the boat will end up on land. If a hatch cover is turned upside-down, the boat will capsize. Whistling on deck will bring on a storm. It's bad luck to throw eggshells overboard without crushing them first. Land birds on boats forewarn disaster.

These kinds of beliefs grow inside me like sea grass on the boat's hull, mingling with what I know of seamanship. If someone asks me how my engine's been running, sweat breaks out in my armpits, and I opt for a roundabout, "Oh, so far, so good." I don't pull out the camera or recording gear when I first spot whales. I wait until I'm with them. If I assume too much, they'll disappear. In rough conditions, when Dall's porpoises surround the boat and ride the bow wave, I feel relief, knowing I'll be safe. I wear the lucky sweater my mother knitted out of smoky sea colors for my first summer on the project. Lance and his wife throw yellow M&Ms overboard when they can't find whales.

Once, two crows landed on the bow of the *Lucky Star* when we were heading out to sea from a harbor. *Oh, no*, I thought, and one of our volunteers, a woman from Australia, looked knowingly at me and said, "Someone's trying to get in touch with you." It can't be helped. Fate is unpredictable, and at sea, the fate of others becomes my own. Every summer, I hear at least one Mayday call on the radio: boats on fire, boats sinking in minutes, boats exploding. Whenever I hear these transmissions, I cry.

I curl up on the damp stones inside the wheelhouse and think of *Rocinante*. I imagine the Gulf a graveyard of ships. Harold has accepted his fate, has moved on. When he told us about the sinking, we were having dinner aboard the *Lucky Star*. He said he lost *Rocinante* because he didn't pay enough attention to the weather. The forecast was off, but fishermen out there don't trust the forecasts anyway. *Rocinante* was uninsured. He had to reoutfit the new boat bit by bit. Though Harold was pragmatic, pleased with his new boat, every time we mentioned a particular instrument or piece of equipment, he wryly said: “I wish I had a barometer.” Or, “I wish I had binoculars.” And he told us: “I dreamed I was off the California coast on a boat, and I saw her. *Rocinante* was up on land, high up on the beach, but she wasn’t mine. Someone else had claimed her.”

The ocean has *Rocinante*, but she doesn’t have Harold. No matter how much a person knows, the worst can happen. No matter how little she knows, luck can spare her.

The way I’ve learned about being on the water is the way I learned to swim. One day in college, I went to the pool and just began. Between laps, I watched the other swimmers. The first piece of advice someone gave me was that I had to keep my head in the water.

On boats, I listen to how people make decisions. I read diesel manuals. I watch mechanics do their repairs. I even ignore their biases for a fleck of wisdom. One mechanic lectured me as I watched him check valve clearances. “Men are different than girls with this stuff. Men aren’t afraid to just take things apart without worrying about getting them back together. You have to just plunge in and do it and not sit around worrying.”

For most of each summer, I’m the one running the boat. During my fourth summer, a woman fifteen years my senior helped me. She grew up in Alaska, had built her own log cabin, had done solo winter dogsled trips in the Arctic. She’d brought all her own gear, her aging husky, even her own wall tent, which she pitched down the beach from the camp. She lit a fire in her woodstove each morning to cook a cast iron panful of canned bacon and eggs.

She’d never spent time on boats and didn’t trust me. In spite of whatever skewed combination of wisdom and foolishness I possessed, I was responsible for the boat, the project, the camp we lived in, for our safety.

For two weeks, we couldn’t find killer whales. At night, after filling out logs, I read Richard Nelson’s *The Island Within*. He wrote about a stretch of bad luck when he’d been unable to kill a deer for his family’s winter food. One of his Koyukon teachers, Grandpa William, told him, “A good hunter . . . that’s somebody the animals *come* to. But if you lose your luck with a certain kind of animal—maybe you talk wrong about it or don’t treat it with respect—then for a while you won’t get any, no matter how hard you try.” Cleverness and skill are irrelevant then.

During that time, I slept two or three hours a night, writing incessantly in my diary, reading Nelson's book. I was convinced that my field assistant's dislike of me, our inability to find whales, were the direct result of my failings.

Finally one morning a group of killer whales passed our camp. As we followed them down the passage in the *Whale 1*, the afternoon sea breeze blew up. The whales left the calm passage and entered Montague Strait, where three-foot seas white-capped the water. Though we'd photographed all of the animals, I hadn't had a chance to record their calls. I wanted to keep following them, not knowing when we'd see them again.

S—— fumed at me. "We have no business out in Montague Strait in this boat. There are big rollers out there. You don't have enough experience. If you want to go out there, you'll have to drop me off on the beach and pick me up when you're done."

"It upsets me that you don't trust my judgment. I've been out in Montague Strait in these conditions before. I've run boats for three summers. I have experience."

"That's not the kind of experience I mean," she shot back.

I turned the boat around, not because I thought S—— was right, but because in wilderness, in groups, the person least comfortable with a decision has the ultimate word. Still, her statement about experience has troubled me all these years, and now I know what she meant.

With a lifetime's experience, maybe even a genetic affinity for seafaring, Harold watched his luck fail. The ocean is merciful only on its own terms. Some careless—but lucky—mariners I know avert every mishap. Around galley tables, they happily relate their tales of near misses over mugs of whiskey-laced cocoa. I think of the ocean's surface pricked through with tiny lights. When mistakes intersect with a light, we go down. Mostly, we're forgiven. Fate is cagey.

It's afternoon now, and the wind has died down in Resurrection Bay. The forecast is the same, but the storm delayed. We try to leave again. This time, there's only a seven-foot swell with chop when we round Cape Resurrection and head east into the Gulf. Gray whales in their spring migration from Mexico to the Bering Sea stream past. Their barnacle-encrusted bodies rise and fall in the swells near the boat, the waves washing over them, the wind blowing their breaths back.

After a few miles, Craig begins to fret. The seas in the Gulf aren't big, but they come from different directions, and the *Lucky Star*'s autopilot doesn't work in these conditions. We'll have to take turns steering all the way across, arms quickly tiring from correcting course.

Night is coming on, and Craig is anxious about the storm blowing up halfway across in the dark. Lance is more optimistic. "We may get bashed around a bit, but we'll make it," he says. But Craig turns the boat around, and again, we head back to shelter in Resurrection Bay.

That night, Craig tells us about the near-sinking of the *Lucky Star*. A couple of years ago, he and his fishing crew were anchored up, everyone asleep, exhausted from the day's work. Hours before dawn, Craig bolted awake.

"Something felt weird," he says. "When I fell asleep, there was a slight chop rocking the boat. At some point, the motion changed. That's what woke me up. The *Lucky Star* felt like she was lurching in wet cement."

Instinctively, he knew what was wrong. When he pulled the hatches off, water slopped over the floorboards.

"My first thought was, *It's done. How the hell am I going to get everyone off of here?*"

The boat's batteries, starter, and alternator were submerged. In a moment of desperation, Craig turned the ignition switch, and the engine roared. The pumps came on.

"How did the water get in?" I ask, feeling sick to my stomach at the thought of the boat's fullness, its wallowing. In another minute or two, it might have sunk, as Craig says, "like a stone."

"With the hold tanked down with fish, we were low in the water. The rocking caused water to lap into the through-hull fittings on the side. The high water alarms weren't working, and I knew it. We were just damn lucky."

I wake in the 5:00 a.m. darkness. Like Craig, I'm an early riser, and on boats, my senses are alert, even in sleep. I look up from my bunk. Craig sits at the helm. He switches on the weather channel and stares up at the display as the broadcast is read. Storm warning, again, northeast winds to fifty knots, seas to twenty-six feet.

Craig turns his head and looks out the window. I know he can't see anything except the reflected lights of the instruments. The weather forecast goes around again and a third time. Perhaps he hopes he's heard it wrong or that on another repetition it will be different. His body looks tense, elbow propped on one knee, chin held in his hand, long fingers curled across his mouth, eyes wide. I watch him closely, reading his thoughts. We've lost two precious days already, waiting here in Resurrection Bay. When his hand reaches for the ignition switch, I blurt, "So what are you going to do?" He glances down at me, and his hand quickly falls off the key as if it were guilty of something. I imagine myself a little devil on Craig's shoulder, coaxing his doubt.

"I don't know," he replies.

Now Lance is awake. "If you go, you'll have to leave me in Seward," I say. Fear is stronger in me than pride at this moment.

"What do you think, Lance?" Craig asks.

"Well, like I said, we'd probably get banged around out there if it came up. It would be pretty unpleasant." His words convey inconvenience and discomfort, not a sense of danger.

Lance is cavalier about this, he doesn't realize how squirrelly the Lucky Star *is, with its shallow draft hull, not the deep V of an ocean-going boat like his,* I think. Panic has replaced trust in me.

We don't leave. Craig and Lance go back to sleep, and I curl up in my sleeping bag and write. *I won't second-guess this decision about when to cross again, because I know both Craig and Lance are cautious and respectful of the sea and experienced. And they trust this boat.* But I'm not at all sure what's right. Lulled by the sound of Lance's deep breathing, I fall back asleep.

At 8:00 a.m., the light wakes me. I glance over at Lance, and he's awake too, reading in his bunk. "I feel bad about being such a naysayer, about being scared of the crossing," I confess.

"Don't feel bad. It was a hard decision. Your input is as important as anyone else's."

I ask him how his wife, an experienced mariner and the most cautious of all of us, would have handled the decision. He chuckles. "Well, she would have listened to the forecast on the hour and taken down weather information in a notebook, noting wind speeds, directions at different stations, changes in barometric pressure. She's scientific about it. I rely more on gut feelings."

Before embarking on careers as biologists, Lance and his wife were lighthouse keepers in British Columbia and made daily weather reports to the Coast Guard. When his wife works with me on the boat, our field notebook is filled with encoded weather data, the habit having stayed with her.

My ideal is somewhere in between, a bit of intuitive sense, a bit of science. In the calmer waters of the Sound, I have a clear sense of my limitations and those of the boat. The open ocean is still unknown to me.

We wait out the morning, eating pancakes and reading. When I step onto the deck in bare feet, they sink in fresh snow coating the fiberglass surface. The air smells raw, like wet metal. After lunch and a tromp on shore, Craig drags a buoy behind the kayak, and Lance and I take turns shooting biopsy darts at it from the back deck of the *Lucky Star*. The sun breaks through the clouds. Out in Resurrection Bay, the water's calm. Again, no sign of the forecasted storm.

At 4:00 p.m., we haul anchor and head for the cape. Above us, the sky is mainly clear. This time, I sense, we're really leaving. I sit up on the flying bridge and write in my diary. *I'm riding into the consequence of my faith and the end of this period of waiting and uncertainty, heading out ahead of the storm. I feel ready for this crossing's disclosure. Hard to imagine a storm now, with the sunlight crackling the water's surface, gleaming through a high, thin layer of cloud.*

The wind at our backs this time, we head once more out to the Gulf, leaving Thumb Cove, with its eroding snow slopes, its shelter, behind. Before we left, we listened to the updated forecast: storm warning, northeast winds fifty knots tonight, seas twenty-six feet. Lance's watch showed the barometric pressure at 998, a drop since this morning. Prince William Sound is six hours away.

As we near the cape, I watch waves breaking over notorious Mary's Rock, north of the point, named for one of many fishing boats that's gone aground there. Due south, beyond a slightly curved demarcation of horizon, the next landfall is Hawaii. I think of Harold making that trip alone on *Rocinante*. On the ocean, I've never been out of sight of land. I cup my hands around my face and try to imagine that strangeness.

A slight darkening of blue, like a shadow on a flower's petal, suggests the storm, but the sea here is calm, with only a slight swell and cat's paws scratching the water where the breeze touches down. Lance joins me on the bridge, glassing the water with his binoculars. "I like ocean swells," he says, "I find them comforting." I can see what he means. And despite the darkness of the front, the sky still looks like something to put faith into. High, opal clouds give way to hazy blue. Yet there is an unsettling quality to the milky light.

The wind's cold teeth chill the back of my neck. Abeam Cape Resurrection, the gray whales pass in small herds and kittiwakes twirl off the cliff face, fluttering, each bird a prayer for sailors, I think. One by one, they land on the water in strings of white dots, like script, like a language I could learn.

There's a calm acceptance that hasn't been there any other day. Was my anxiety a premonition or just worry? I think of the mechanic's advice, to just dive in and do it instead of sitting around worrying. His advice is one voice inside me; another is intuitive. For now, the intuitive voice speaks louder.

At 4:30, we're out in the Gulf, several miles offshore of what one book calls "a stern and rock-bound coast." Another book title describes it as the place "where the sea breaks its back." Though deep glacial fjords penetrate the coast, they are open to the south swell, offering no protection. Off Day Harbor, we enter a wind line. I've been watching it edge closer, a dark, expanding ribbon. Breezes scuff the sea's surface, and a strip of cloud widens out from the horizon. "That front is coming closer fast," Lance says. The moon is up, waxing, translucent, the front's thin edge gauzed across it. We watch birds through our binoculars, scan for whales.

At seven, we approach Cape Junken, halfway across the Gulf. I'm alone on the bridge. I shut my diary. Every part of me is stiff with cold. Though the sky is still blue above us, clouds have pushed their fingers into the Sound. Blowing from the east, wind whitens the water. The swell rises into gray brooding hills.

I go below into the cabin and sit at the table across from Lance. He sets his novel down, pulls off his headphones, and glances over at Craig, then out the window. A few seconds later, Craig looks up from his reading and out as well. Some invisible signal, the sea cresting a particular height, alerts each one of us simultaneously.

Craig reaches for the logbook and scrawls something down. I pick it up to see what he's written: *Off Cape Junken, swell 10 feet, wind rising, the storm moves in.* Lance and I scan the chart, estimate that we're two hours from Prince William Sound. "We'll get a little wind here at the end to spice things up," Lance says brightly, putting his headphones back on and turning back to his book.

As we pass Cape Junken, the wind and swells continue to build. A dark-hulled limit-seiner, bigger and more seaworthy than *Lucky Star*, heads west, passing within a few hundred yards of us. It sinks, disappears, rises, riding the swells. Craig jumps up from his seat, looks out the port window, and exclaims, "It's the *Lady Luck*! That's Matt—Matt Luck." He grabs the VHF mike.

Matt Luck describes the herring fleet waiting in Cordova for the forecast to improve before heading to Kodiak for the fishery opening. A log barge has been circling the Sound, he says, day after day. Now some of them no longer believe the forecast and, like us, are heading out. But unlike us, these boats have a thirty-six-hour run to Kodiak. I think of the *Lady Luck* heading into what mariners call the teeth of the storm.

The growing swells make me seasick. I crawl into a bunk and doze as the *Lucky Star* rumbles and rolls. We've nearly crossed the Gulf now. We're committed. There's no shelter to duck into between here and the Sound. The storm is here, and soon the Gulf will take on the character of its name, a chasm wide enough to swallow a ship, a life.

The last time I saw *Rocinante*, she was anchored in a cove in the Sound. I rowed out from Whale Camp to visit Harold in the evening. The water that night was almost black from reflected trees, silver light still gleaming near shore. As I nudged *Rocinante*'s side with my inflatable raft, Harold put down the fish he was baiting, wiped his hands on his blue coveralls, and reached to take my line. I hauled myself aboard over the *Rocinante*'s rail. Harold had carved its supports elaborately, like a banister. He gave me a strong hug. "Good to see you!" he said, with his faint Norwegian accent.

I offered to help him bait his hooks for the next day's fishing, which would begin at 5:00 a.m. The deck was slick with herring slime. Harold showed me how to impale the herring on circular hooks, and we stood on either side of the white plastic tote that held his catch. The herring left gold scales on my hands. I had to grip the stiff, slippery bodies tightly, my fingers clumsy from the cold. Harold told me about the new electrical system on the boat, about his catch, that it had been a good day. I told him about the whales we'd seen. Then we were silent.

When we finished baiting, he lifted the tote cover and showed me his cleaned and gutted catch. The black cod lay smooth and iridescent against one another. I imagined his hands moving over each one, smoothing away the slime, scraping out the entrails. I imagined him unhooking each fish after the long wait while the line "soaked" on the ocean floor. How each fish seemed beyond value. Now the *Rocinante* lies where those black cod lived. Some might even find shelter in her burst-open hold.

A shout wakes me up. "Let her go!" The anchor clatters. We're in Foxfarm Bay, just inside the Sound. I think of a line from *Don Quixote*: *If luck had not caused Rocinante to stumble and fall midway, it would have gone hard with the daring trader*. I don't know how fate works, why *Rocinante* sank and Harold survived. Perhaps that storm saved Harold in more than the obvious way. Whenever I see him or write him a letter, I wish him safe passage.

If fate is cagey, then so is practicality. I don't know Harold's heart, only what he tells me. Once, coming back across the Gulf with him after a fishing trip, he looked out the window of the new *Rocinante*, glanced at his navigation plotter, then at me and said, "This is where my boat is."

In the silence after the engine's been killed, I hear water lapping the hull. I think how we're caught in different currents, Harold and me, Lance and Craig, Matt Luck, the herring fleet, all held within the ocean's bestowal. At sea, we're more the same than different. This is how its power is binding, how, by water, we're joined.

The next day, safely anchored up as the storm busts over the Sound, we listen to mariners chat on the radio. We hope for killer whale sightings. Instead, we hear skippers swapping storm tales. One man says he saw the weather map, the clotted cloud mass like the satellite picture of a hurricane. Another says he crossed the Gulf a couple hours after we had, the night before. "There were twenty-foot swells off Cape Junken," he said.

I say a prayer for the *Lady Luck*.

Wondering Where the Whales Are

But what surface have we fallen through,
Here beneath the trees? What do we see in our infinity
if each is all the same, or all unknown?

—Molly Lou Freeman

Halfway down an alder slope, through an opening in the jumble of branches and leaves, Craig's son Lars spots them. "Look, killer whales! They're coming into the bay!" Even though he's only eight, Lars knows whales from a mile away and five hundred feet up. Since he was six months old, he and his sisters have been bundled and stashed on his parents' research boats every summer. At Whale Camp, he rubbed holes in his baby socks bouncing on the wall tent's plywood floor, his Johnny Jump-Up suspended from the ridgepole. "Listen," he says, "you can even *hear* them. They're over *there.*" He points; we squint. There, in ponds of sunlight, we see black fins, breath smoke.

"Good eyes, Lars," says Craig. "Let's go." We lurch down the mountainside, branches tearing at our hair and our faces.

On the beach, we gallop, digging wedges into sand with our rubber boots. In the distance, a few yards offshore, the first bubble-cloud rises, and we run for it. A fin slits shadows at the sea edge. When we get closer, we tiptoe. Five killer whales slide along shore, releasing air so they can sink. White bubble-rings bloom and phosphoresce on the water's surface. It's called beach-rubbing. On beaches with particular slopes and small, round pebbles, killer whales approach shore to slide their bodies along the bottom. One bay to the south, in the winter of 1918–19, the artist Rockwell Kent, holed up with his nine-year-old son at the homestead of Lars Olson, a seventy-one-year-old Swede, observed the phenomenon. Kent might have seen the mother of Aialik, the oldest female in this group, rubbing on his beach.

Craig and I, unthinking, pull off our boots and socks, make to unbutton our shirts and jeans, to jump into the water, where the whales are. Lars asks, "What are you guys doing?" and we look at each other and laugh. What *are* we doing? We fall heavily onto the beach, side by side and watch. The fifty yards separating the whales from us might as well be the Gulf of Alaska. We can't—and shouldn't—cross it.

Even though we've been studying killer whales for two decades, fishermen still don't quite understand what we do out here. They call us "whale watchers." They shake their heads at Craig, a fisherman himself for twenty years, who, instead of sleeping during closures, raced off in his seine skiff to find killer whales.

Listen, I want to say, we're not *tourists*. We're doing *research*.

But what we do is watch. We watch from shore, with our boots askew on the ground. We watch from the boat's deck, poised with our notebooks, pencils, cameras, binoculars, with vials in which we place samples of whale skin. The whales visit our dreams, where they watch us. So what should we be called—scientists, voyeurs, observers, natural historians, writers, intruders, watchers? The killer whales are called *aaxlu, takxukuak, agliuk, mesungesak, polossatik, skana, keet*, feared one, grampus, blackfish, orca, big-fin, fat-chopper. Whale killer. From the realm of the dead. *Orcinus orca*.

In grad school, listening to a lecture, I stare out the window and scribble along the margins of my notebook pages:

outside
birch fingers cast smoke
ribbons on snow

The professor chalks formulas on the board, flips on the overhead projector's light, casts a graph on the wall of the oxygen consumption of a marine mammal. *The language is like this*, I write in my notebook:

kinetic
isotope
extrapolate
index

The formula makes sense. Someone has figured this out, disproved an old theory about the way whales dive. I memorize the formula, stash it like some possibly useful thing into that catch-all drawer in my mind.

In another part of my mind, in a dream, I'm standing on the *Lucky Star*'s stern with Lars. In the boat's wake, a pod of beaked whales circles. They twist in the foam, their eyes glinting up. When Lars reaches out his hand, a whale grabs it and won't let go. Using all of my strength, I yank him free.

When I told a Yup'iq woman about this dream, she called it a warning. In her Bering Sea village, she was taught that killer whales must never be hunted or bothered, that we mustn't touch them.

But we want to know.

We want to know, like the raven, who, in a Chugachmiut Native legend, swims into a killer whale's mouth and finds an old woman sitting inside the whale. She asks, "How did you come here?" The raven answers, "I called him. I wanted to know what was inside him."

During my first summer out in Prince William Sound as a volunteer, running the *Whale 1*'s predecessor, a tiny, problematic skiff my field assistant and I finally—after countless breakdowns—dubbed "Dorky Orky," one of my tasks was to decide on a project for my master's thesis. Initially, I felt drawn to the quieter ways of humpback whales, who stayed in protected areas near Whale Camp to feed. But small groups of killer whales kept passing by camp, hugging the shoreline. They were AT1 transients, mammal-eaters about which little was known except that they were mostly silent and difficult to follow. But we did follow them, into and out of every bay and small passage in the southwestern Sound as they searched for harbor seals. Sometimes, anchored up in a tiny cove, we woke in darkness to their sharp exhalations around the boat. Other times, after following them for hours along shore, one of us peering over the bow to watch for bottom, the whales would veer into open water and attack Dall's porpoise pods, throwing the 400-pound black-and-white animals—resembling miniature killer whales—twenty feet into the air. My mantra became, "If we follow them long enough, we'll see something amazing." I grew curious about how those mostly silent whales communicated.

One day, my friend and I followed two AT1 transients from a small inflatable as they hunted harbor seals along an island shore. We lost them for several minutes, and then spotted silver mist above a rock. We let the boat drift near. Clinging tightly to the rock, its head craned back, eyes huge and black, a seal pup crouched above the waterline. A transient nudged the rock, but couldn't reach the seal, at least not yet; the tide was rising. Abruptly, the whale turned, joined the second whale, and swam rapidly across an open passage. We left the lucky seal and raced to catch the transients, but they'd vanished. Cutting the outboard in mid-passage so we might hear their blows, we stood up, scanning with binoculars.

I felt something through the bottom of my feet before I heard it. From the inflatable's wooden floorboards, a wail rose, and another, and another. My friend and I stared at each other.

"It's the whales. They must be right under us. Let's drop the hydrophone," I said.

I scrambled for the tape recorder, and we huddled over the small speaker adjusting knobs as long, descending, siren-like cries reverberated against underwater island walls. In the distance, other whales answered, faintly. I'd never heard transients call before. It was like a stone had sung. I knew then. I wanted to learn the language of the whales that were mostly silent.

In grad school, I learned the art of detachment, learned to watch how I said things, to listen for anthropomorphism, like applying the word *language* to non-humans. As scientists, we distinguish ourselves from whale huggers, lovers, groupies, and gurus, from those who think of whales as spiritual beings. We learn the evolutionary, biological basis for an animal's behavior. We study the various theories and counter-theories and debate their merits: reciprocal altruism, game theory, optimality theory, cost-benefit analysis. At scientific meetings, in animal behavior seminars, we don't debate whether animals have feelings. It's *terra incognita*. But on the research boat, or at the breakfast table before the meeting begins, some of us talk about these things. One non-scientist friend, puzzled by the ways of science, asked "Isn't it strange to assume that humans are the only creatures with feelings, that we are so different from other animals?" Is it "animapomorphic" to ascribe animal traits to humans? If it's wrong to suppose that animals might share qualities with humans, then how do we see ourselves? Alone at the tip of some renegade branch of the tree of life?

Out in the field, summer after summer, we search for knowledge, employing the scientific method: observation, hypothesis, data collection, analysis, discussion, conclusion. Poet and biologist Forrest Gander says that this method "has endured as a scientific model, and a very successful one, for it predicts that when we do something, we will obtain certain results. But if we approach with a different model, we will ask different questions." To create a new model: that prospect challenges all of the questions I've learned to ask—and not to ask.

⁂

Over the course of a four-month-long field season, sometimes we see killer whales every day, and sometimes weeks go by without them. Often, we've spotted distant whales, come near them to take photographs, and they've vanished. Other days, we've followed killer whales for twenty-four hours. Then, after so long, even the observing eye becomes insufficient, so we listen. In darkness, we navigate by the sound of their breathing.

⁂

We take turns sleeping. I'm leaning on the boat's dashboard, echo of engine roar dying, wave slaps against the boat's hull taking over. I drop the hydrophone down to listen. Two in the morning, just past summer solstice, Montague Strait is dimly lit, but it's too dark to see the whales. I don't hear anything. They're down for a long dive, or we've lost them. I hold my breath. Then, a few hundred yards away,

whoosh-ah   *whoosh*   *whoosh whoosh-ah*

I stare in the direction of the sound, hear water closing around it.

Another summer. For eleven days, Molly Lou and I search hundreds of miles without finding killer whales. On the twelfth day, we hear radio reports of a large group forty miles from camp, in open water. After roaring past Smith Island and Little Smith Island at twenty knots, we drop the hydrophone, climb onto the cabin top, scan with binoculars, radio the boats who reported the whales. The water's still. No one responds. On the hydrophone, waves lap, lap, lap.

After searching an hour, we give up, then devise a plan. No more running around wasting fuel. We're going to wait, let the whales come to us. We return to camp, and the next morning we gather paper and books, food and a thermos. We run the boat a mile off camp, out in the passage, and shut down. After dropping the hydrophone and scanning around, putting out a radio call, we settle into the boat, build a fire in the tiny stove. Every half hour, one of us pulls on raingear and climbs onto the roof to scan with binoculars.

A couple of hours pass this way. I look at my watch, put down my book. It's my turn to scan. I glance out as I reach for my jacket. Fins rise around the boat.

It worked! I shout, and we rush to put on raingear, gather cameras. The rest of the day, we follow the whales.

The same trick never works again. We're constantly second-guessing. Should we sit still? We call that the "sit and wait" hypothesis. Should we move? We call that the "Lance and Kathy" method, after colleagues who averaged a hundred miles a day one summer. Once, Craig and I searched the outer coast of Montague Island, over seventy miles, and saw nothing. Our friend radioed that killer whales were off Point Helen, a few miles from Whale Camp, and only about ten miles from us, as a raven would fly. If we could have, like the lunatic film character Fitzcaraldo, drug our boat over that mountain, we'd have been there.

That night, we watched a wildlife DVD about a group of filmmakers who spent four years trying to photograph snow leopards in the Indian Himalaya. They never saw a kill, never saw a litter of cubs, their two greatest desires. They didn't see snow leopards for the first several weeks, just tracks and scat. They concentrated on these tidbits, mapping them until they sensed patterns. Even after decades in the field, we constantly have to revamp our intentions and strategies, remind ourselves to concentrate, not on our desires, not on the past, but on clues, which is hard when you're lifting fluke-prints off water.

The morning after we watched the film, three AT1 transients slid past our anchorage. We heard them first on the hydrophone. They were hunting marine mammals offshore, diving for ten minutes at a time, constantly changing direction. The north wind blew up. We lost them after an hour. We managed to take one identification photo.

In the Tlingit language, the word for killer whale, *keet,* means "supernatural being." We'll never know its true connotation, but it fits. In nature, creatures defy our assumptions. In the 1980s, biologists divided fish-eating killer whales into pods, extended family groups that remained together for life. Recently, that story has been revised. These societies orbit around the matriline, mothers and offspring. Pods can fracture. The loss of a key female may cause a family to rupture, for bonds to loosen. Discoveries reveal the *keet* nature of the wild animal. And the more we know, the longer we stay, the more we care, and caring, like anthropomorphism, is tricky ground for that detached creature, the scientist.

For the past few years, we've been collecting samples from killer whales to measure contaminant levels in their blubber, to extract DNA from their skin. We've learned that their populations are small, a few hundred animals, so an oil spill or a die-off of salmon or seals can be catastrophic. We've confirmed that residents and transients don't interbreed, though they share the same waters, that transients carry high PCB and DDT levels in their blubber, that mothers pass these poisons to calves through their milk. But to learn this, we have to approach whales more closely than we do to take photographs. To do this, we point a rifle at a whale and shoot a biopsy dart into its body. The dart pops out after snagging an inch-long piece of flesh on its thread-like barb, and we scoop it from the water with a dip net. To do this, Craig and I argue through our conflicted feelings. *We can't dart now; they're resting. These animals are rare. We can't dart in front of tour boats. We might not have another chance. We've probably darted enough animals in this group. We need more samples for the statistical tests. We have to have a common mind. I hate all this.*

Even Lars, who's enthusiastic about shooting, scrunches down in the bow, fingers plugged in his ears, eyes shut tight when the shot's fired.

From the boat's cabin top, I scanned Montague Strait in light diffused by high clouds, looking for blows. I spotted a white glittering, then another. It was the kind of haze made by a leaping whale when its body collapsed onto the water.

We raced that way and found killer whales, took identification pictures of their dorsal fins and flanks, recognized them as Gulf of Alaska transients. The last time I'd seen them was four years before. They'd never been biopsied, but we knew that their calls differed from those of the local AT1 transients, so they might be from a completely separate population. That day, Craig wasn't there to wield the dart gun, and my field assistant—my husband, John—and I had to do it ourselves.

For the next two hours, the whales led us past Danger Island, into the Gulf. John, more comfortable with a rifle than I was from his years in the Alaskan bush, shot three times without success. Out of Montague Strait's strong current, the water calmed to a swell. In my impatience, I took the gun. John pulled the boat in close to the whales, and I sighted on an old female's scarred saddle patch. Without thinking, I pulled the trigger. The dart hit the saddle patch and bounced out. She slapped her tail and dove.

"We got a sample," I shouted, elated, when I pulled up the dart and saw blubber protruding from the tip. I gave John the gun on the next approach, and he darted another female.

"We're getting pretty far out here," he said after I wrapped the third sample in foil and stored it in the cooler. "I think we should go back." I glanced toward the Sound. We were at least four miles from shore now, and the whales were heading steadily south in the direction of Hawaii. As we drifted, we watched them disappear.

An hour later, anchored up at Foxfarm Bay, just inside Cape Elrington, intent on processing samples and thrilled at our success, I didn't notice John watching me.

"I've never seen you that way before," he said.

"What way?" I asked, looking up.

"You were so angry and impatient, even rude at times, and then, suddenly, when you got what you wanted, you were ecstatic. A real Dr. Jekyll/Mr. Hyde thing. It was scary."

I stared across the bay, where a sea otter lazily rolled and dove and brought up some kind of shellfish. Inside me, a nauseous feeling rose.

I haven't darted many killer whales since. It's Craig who wields the gun. And there are whales we've never been able to dart, mostly sea lion hunters with torn fins. They sometimes approach our boat, curious, staring at us with inscrutable eyes. Once, a female grazed her body along the skiff's side, her mouth open, showing rows of perfect teeth. "What are you saying?" I called after her as she swam away.

Years ago, another whale drifted under the bow where I stood, looking down. She held a harbor seal in her jaws. Blood from the seal's body throbbed.

Science trains me to be detached in moments like those, but sometimes I'm angry or panicked in the field, when I can't get what I want, what I *must* have. When I face the fact that I have no control over what's invisible, what binds me so viscerally to my desires, what decides when the whales will find me.

After several days without whales in Resurrection Bay, Craig and I overhear a radio conversation between tour boats. Killer whales are traveling along the rocky shoreline of Fox Island, fifteen miles from where we're floating, our hydrophone down. They're heading for the cape, out of the Bay and out to sea. The skippers think they're transients—the ones they call "the bad boys"—two large AT1 males that hunt harbor seals in ice floes off the Aialik Glacier.

We drop our books and scramble to start the engine, call a skipper, get a location and direction of travel, and roar across the Bay, coaxing as much speed as we can out of the *Whale 2*. When we spot the whales, we know right away they're not the local "bad boys." Their fins are too broad and tall. As I slide the boat in parallel to the whales so we can take pictures, I scan photos of transient dorsal fins in the killer whale catalog.

"Who do you think they are?" Craig asks, clicking off frames. "They're awfully tolerant for Gulf of Alaska transients."

The whales travel slowly, breathing for eight breaths, then diving for ten minutes. They follow a regular compass heading east, directly past Cape Resurrection, toward the Sound. I stare at two blurry photos, then back up at the whales.

"They *are* Gulf of Alaska transients. They're the AT30s." The pictures are poor, taken during a single encounter seven years ago in bad weather.

We spend the next hour trying to get biopsy samples. Tour boats come to watch them, so we don't dart. Darts miss. Once, a dart pops out of a whale but doesn't take a sample. Another time, we're too far away when they surface. Other times, they change direction slightly when they dive. I plead to them, to Craig's amusement, as I position the boat. "Whales, please let us take these tiny samples. We'll never have to do this again. It's for your own good!"

We call out names for them, Chubby Rain and Heavy Rain. Despite our blundering, our absurd behavior, the whales let us approach closely again and again, and finally we have some samples.

Floating off Killer Bay, we watch them disappear. "Don't you wish you knew where they were going?" Craig asks. "Someday, with a little transmitter attached to them, we won't have to wonder where they are."

Now I can barely make out two distant black triangles among rolling hills of water, and I think of them unwatched by anyone for eight more years. They're swimming off the edge of the known world, like hapless ships on ancient charts. They might dive right through the sea realm, resurface in some other, a realm of the supernatural. A young Sugpiaq man from

Nanwalek, a tiny village in outer Cook Inlet, told me there's a lake near his home that's bottomless. A killer whale jumped into that lake, he said, dove to the bottom, pushed through and emerged in another lake.

We cling to what we know. In response to Descartes' mechanistic view of the universe, Blaise Pascal said, "The silence of these infinite spaces terrifies me."

Science. It seems solid, but it's mostly space, like a gill net I drop over the world. Two transients pass through its web, leave me holding a tiny sample, a pencil.

A young scientist seeks mentors. Bud Fay, my major professor in grad school, an expert on the walrus, showed me how a scientist could learn from and gain the respect of Native people. Hunters on St. Lawrence Island still remember him. Craig, other whale biologists, and those I know through their discoveries, their tenacity, their eyes that see and ears that hear what others miss, are my biologist heroes. I met Mike, my last mentor, one afternoon at Chenega Village. He rode his four-wheeler down the steep ramp to the dock. In the vibrating silence after he'd shut down the engine, he sat and watched me as I pumped fuel onto the *Whale 1*. His look was inscrutable. There was no smile. Under his cap, his eyes were shadowed. He could have been angry. Non-Natives were not always welcomed in the village. I tensed when he climbed off the four-wheeler and, hands in pockets, strolled over to the boat. "Seen any whales?" he asked, grinning.

He was all sinew, brown skin, black hair, and a small, bowlegged frame. He wore a plaid wool shirt, stiff new dungarees, and wire-framed glasses. I knew he was considered a village elder, although I couldn't tell his age. He coughed often, into his fist, turning his head away. I introduced myself, but afterward, he'd show up at the dock whenever I was there and greet me, "Hey, *Whale 1*."

He dropped bits of knowledge into our conversations, where he'd seen whales, how seals in the area were declining. I knew he hunted seals and fished for salmon but learned only from other villagers that he was one of the most respected elders in the Sound and one of the last seal hunters in his village. I also learned he was dying of lung cancer. He'd gained his knowledge by roaming the Sound in a boat in all seasons, watching. Since the oil spill, he'd assisted biologists on their projects—on octopus, harbor seals, subsistence traditions—and strove to involve his village in the science.

I began to look for Mike when I came to Chenega, wandering to his house, inviting myself in for a cup of tea. Somehow, I felt attached to him. Our conversations were brief. But, after time, when he saw me, he hugged me. He teased me. When I told him what I wanted to be, he shook his head. "Why do you need to do that? You don't need to go to school to do that. You just need to live out here."

The smell of burning alder drifts up from Mike's smokehouse. He's gone today. He's hunting seals.

Molly Lou and I anchor the boat in front of camp. It's sunny, but the wind's come up, so we decide to take turns trying on the dry suit, snorkel, and mask and swimming through the eelgrass and kelp beds. Molly Lou helps me with the zipper.

I put on
the body armor of black rubber
the absurd flippers
the grave and awkward mask.

I hear words from Adrienne Rich's poem in my head when I drop feet-first from the boat's side into the sea.

There is no one
to tell me when the ocean
will begin.

After I pull the black rubber away from my neck to release air, the dry suit clings to my body like loose skin. I place my face in the water and breathe through the snorkel, wheezing rapidly at first out of fear, and the sound is loud, like the breaths of someone dying.

Eelgrass and kelp stream below me. Now my breathing sounds as if someone is breathing for me. I paddle. I make arcs through the water with my hands. Tiny sculpins wink in and out of battered fronds. As I swim along a rock outcrop, I look for seals. I glide along rocks and quiet my movements, searching the sandy bottom. My body blots out the light above me. I'm hungry. I search the whole island's submerged perimeter.

At times like these, I get closer to the water.

A friend of mine kayaking in the Sound met Mike once and asked him if he knew me. Mike chuckled, said, "I hear her on the radio . . . She's wondering where the whales are."

Mike died four winters ago. The last time I saw him, he had to breathe from an oxygen bottle.

According to traditional stories from all along Alaska's coastline, when killer whales come into a bay, someone will die. A Sugpiaq woman from Nanwalek told me why. "When killer whales come near the village, they're calling someone to join them, so we're sad. A week or two later, someone dies."

There's a killer whale we've named Jack, after the late Jack Evanoff of Chenega Village. His niece, Mary, told me that when Jack was an old man, "He always said he'd come back as a killer whale with a partially bent over dorsal fin." People told her that they'd seen such a whale out in the Sound. It's true. There's only one. He's a salmon-hunter from a pod that centers its range in Prince William Sound. His fin curls to starboard, and from the back it looks like a question mark. Lars calls him "Captain Hook."

Sometimes, anchored up in a storm in a place called Pony Cove, I joke with Lars about killer whales, make up crazy stories about what they do. I tell him an old tale:

Long ago, a man from Nanwalek followed some killer whales in his kayak. He thought they might lead him to seals. The whales dove at the head of a bay and disappeared. When the man paddled to shore, he saw human footprints leading into a cave. He followed them. Inside, he saw humans putting on killer whale skins. Once, I tell Lars, humans and animals spoke the same language.

Can science teach me this language?

Science teaches me that there's a truth somewhere, that I can find it, that I can listen and hear something. For years, I recorded the sounds of transients. I scrutinized each call on a sonograph analyzer. I scribbled descriptions of everything I saw. I identified hunting calls, resting calls, social calls, long-distance contact calls, but I never deciphered the language of the whale that eats only mammals, that speaks mostly silence. The language of the killer whale eluded me.

What I did learn was that it's not difficult, in the moment, to surrender to not knowing. To be a watcher. Like a transient, who finds its prey by listening, to be silent.

Once I found a picture of Mike as a child on the schoolhouse steps at Old Chenega. I recognized him by his big ears. Sometimes, Mike walked with me along the shore of his island. He'd stop suddenly, motion me to be quiet. "Listen," he'd say. "I hear something."

Montague Island's reflection extends a long way down into the water in the afternoon's heavy light. Whales swim along its snowy flanks, across green slopes, skim the tops of conifer stands, along bare rock, then dive down *into* the mountains. When they rise again, they break apart the island's reflection.

Lars drops a stone into the water. We watch it. The deeper we go, the more knowledge resembles a question mark. Who's asking the questions? We listen. We watch the stone sinking. We watch it spiral out of sight. Science. It's like that.

And Suddenly, Nothing Happened

for Eve

These are the times in life when nothing happens,
but in quietness the soul expands.

—Rockwell Kent

It's a tiny space. Twenty-six feet long, most of that deck. The cabin—eleven feet wide, no bigger than a bathroom—perches on the stern. The wide bow flares out, the sides three feet high, so we can sit on the chipped gray-painted fiberglass and read with our shirts off on sunny days. We live here, a week at a time, sometimes one person, most often two. This time there are three of us, Craig, his oldest daughter, Eve, who turned sixteen a few days ago, and me. Eleven years ago, when Craig bought this boat, then a decayed, waterlogged hulk dry-docked in a muddy field, we changed her name to *Whale 2* so it would be easy for other boaters to remember, to associate the boat with its mission—studying whales. What it seems we do most, though, is *look* for whales, wait for whales.

A soggy batting of gray cloud drapes the mountains today. We're in Resurrection Bay, a fjord shaped by the retreat of a massive glacier. The ocean bottom, a thousand feet deep, still rebounds from the memory of ice, the enormous weight once pressing it down. Ice tongues, another kind of memory, dangle from high ridgetops, aquamarine against gray stone. Pushed upward by the clash of plates at the continental shelf break sixty miles offshore, the mountains are still rising. Along the finger of Cape Resurrection, the pillow basalt landscape is rounded rather than angular, the lumpy, purplish rock formed by rapid cooling millions of years ago, when lava oozed through seeps on the ocean floor.

Through binoculars, I scan the crackled water surface for blows, let my gaze drift up the forested slopes of Cape Resurrection, past tree line to its craggy rock spires until my head falls back against my spine.

"Looking for land whales?" Craig ribs me from his perch on the cabin top. I smile with half of my mouth and drop my binocular gaze to the other shoreline.

The whales should be here now. This is the time—late July—when coho salmon return from the Gulf, homing for their spawning grounds at the bay's head. The schools round Cape Resurrection and Cape Aialik and follow shorelines, pooling up in cup-shaped bays like Pony Cove, Sunny Cove, and Agnes Bay, that gather them like fish traps. Sometimes, a hundred skiffs and pleasure cruisers troll off the points of those bays. Among them horned and tufted puffins, common murres, glaucous-winged gulls, black-legged kittiwakes flicker and shriek. For scraps, these birds shadow the bigger predators—humans, Steller sea lions, and harbor seals, and the animals we seek, the salmon-hunters, resident killer whales.

This is the time when we take our boat out here and wait for whales, and we arrange our notebooks, cameras, and recorders, and fix up the *Whale 2* and make ourselves ready. We assume our shape, and we shape in our minds what will happen. The salmon will come. The whales will come, traveling their familiar routes. We'll see the pods—extended families we've observed for nearly twenty years—among them, the familiars, the "local yokels," like the ADs, the AKs, and the rarer pods, the AXs, the AWs. We'll record each pod's unique vocal dialect. When a whale grabs a salmon at the surface, we'll sieve the water for fish scales with our net to identify what kind they've killed: chinook, coho, humpy, sockeye, dog. We'll photograph the mothers with new calves, and we'll see if anyone's died over the winter. We'll name new calves, and we'll remember and call out the names of ones so familiar, we recognize their fin shapes from hundreds of yards away: *Aialik, Rockwell, Aligo, Phoebe, Aerial, Zephyr, Barwell, Montague, Igagutak, Herring Pete, Josephine, Sather, Keet.*

But nothing happens. The whales aren't here. Killer whales haven't been seen here for four or five days, the tour boat captains tell us. At times like these, I joke that our boat should be called the "Whale None." We float a while in the middle of Resurrection Bay, between Rugged Island and the turbulent current-ripped cape—its Sugpiaq name, *Aialik*, means "the strange place." While Eve reads on her bunk in the cabin, Craig and I sit outside looking through binoculars. Moving black dots stop my scan. I track surf scoters. Their white wing patches flash, open and closed, and the spattering lengthens like a candy thread. Part of the thread disappears behind an ocean swell, which creates a gap, and the line reforms and moves off, trailing questions across the stillness and absence. It leads my eye across a gloaming, windless expanse of sea rising and falling, like the breath of a sleeper, a sea the boat captains call "greasy."

"I hope this isn't going to be a real drag for Eve," Craig says. "She's going to be bored."

"Not necessarily," I say, feigning optimism. "She seems content so far, reading her books." Eve's just returned from a month in Gustavus, the tiny town where she was born, and she had a boyfriend there, her first. She's giggly when she talks about him. He's cute, but kind of wild. She wears a beaded choker that spells out his name, Shreve.

Maybe it's our own boredom we fear. In town, we run or swim to free our loggy limbs, we chat on phones, dig in the garden, make lists of errands and tasks, cross them off, fall on our beds relieved and exhausted. Here, three paces across the deck define our limits. A plunge in the sea wakes us up each morning, but we can't stand the cold for more than a few minutes. The rhythm of water lapping the boat's hull shapes our day.

"Anyways," Craig says, "where are the whales?"

While we wait, we float behind islands, listening to the tour boat operators' VHF radio channel. With a hundred or more tourists on each boat desperate to see killer whales, they'll find them, we know, if they're here. We roam about and glass the water until our eye surfaces ache. We scrub the boat's deck with salt water pulled up in a bucket. We wash the windows. We write, we nap, we tell one another what's happening in the books we're reading. Eve's novel is about four children locked in an attic, a mystery, and we exclaim at the latest horror as if it's a newspaper story. Craig is reading *Tuesdays with Morrie*, a dying man's story, and his eyes get teary when he tells us about Morrie's insights into life and death. I'm reading about poetry, in Jane Hirshfield's book *Nine Gates*, and about intimate places—shells, corners, miniatures, houses, nests, and wardrobes—in Gaston Bachelard's *The Poetics of Space*.

Jane Hirshfield says that creativity "demands that we turn away from our . . . desire to create a shapeliness that does not reflect how awkward, unfinished, and ambivalent actual experience is." I pull out my computer, inspired, begin to write what this is like, this not finding whales. I read sections aloud to Eve and Craig.

But research demands a shape. Data sheets with maps on the back wait to be filled out. Cardboard boxes of black-and-white film and shrink-wrapped cassette tapes fill a drawer in the boat. To be faithful to our long-term study, we must, each summer, photograph the three hundred or so killer whales that roam through this region, a hundred-mile-wide section of the Gulf of Alaska called Blying Sound, encompassing Prince William Sound and the Kenai Fjords. The whales roam over an unknown area, sometimes here every day, sometimes gone for days, weeks, or months, we don't know where or why.

But it's a constant source of speculation. With binoculars pressed to our eyes, facing opposite directions, Craig and I talk. Our voices drift over the water, as if we're chatting to the waves, or to the gull who's plopped down nearby, paddling its feet, looking hopefully at our fishing boat. "The fish are here now," Craig says, "it's a big run." The gull twirls a tight circle. "But maybe there are more fish somewhere else."

"But if it's only food that drives them," I say, "why don't they find a hot spot, like Pony Cove, and park themselves for a month or two, like these fishing boats?" The gull shrugs its wings, picks at the water. Craig suggests that the whales are like Australian aboriginals, going on walkabouts, needing to connect with their landscape, its boundaries, to keep their connection vital. "Good idea," I say.

Perhaps I'm here to do the same, I think later, with my computer humming on my lap, to describe, in detail, what I see. But, as usual, what I see is often blurred by what I wish was happening, and further, by the contents of my head.

I sit on a pink buoy ball on deck and watch as Craig brings us near the shore of Cheval Island to show Eve the seabird nests, the rock ledges and grassy hummocks where they lay their eggs. Craig and Eve stare through binoculars into crevices and caves where nests are hidden. I stare at the water. Here's a pigeon guillemont, with a pure white shield on her wing, floating on the water, where the gray shadows turn almost black. She's dipping and preening. Cormorants fly past, the sky so opaque, they barely form shadows on the water with their wings. "Watch this," Craig says, binoculars pointed up the slope. A horned puffin tilts, then dive-bombs from its perch, wings tucked against its little cigar-body, then flaps, brakes, lands with a white splash. A slight breeze strokes the milky gray water smooth behind it.

As Eve and Craig exclaim, search the cliffs for more birds, I think about the shape of this day, of its many shapes. I'm fogged in my mind. I turn over and over the same flotsam, hoping to find something new. I think about my life, which has assumed, lately, the shape of piled up cardboard cartons, labeled *winter clothes, books, kitchen stuff, fragile stuff.* I don't like shapelessness in my life, but I've just moved south, after thirteen winters in Fairbanks, four hundred miles north of here. I've moved to the coast, to be with Craig, to be closer to my work on the water, to escape the extreme winter darkness and cold of Alaska's interior. I've left a marriage, many friends, my little cabin in the birch woods. I'm anxious to order my life in a new town. I'll need to find a cabin of my own. I'll need to find work. I want to make friends, find a sense of community. I'll need to make space in my life for writing.

The boat's a hiatus. There's nothing I can do here about any of those things. I'm confined to this twenty-six-foot space with my books, binoculars, diary, with the radio chatter, the bounce of waves, the roar of engine, the desire for whales. I try to make myself more compact to fit this space. Privacy becomes a corner of the deck, my eyes closed, not saying anything, or washing the dishes in the plastic pan filled with seawater, or staring at the water with binoculars, looking, wishing, almost but not seeing, a slim, black spear shape rising and falling, a series of misty puffs disappearing—whales. Nothing breaks the water's sheen but boat wakes, which confuse me at times with their black edges and white curls. Distant birds also mimic killer whale fins, rising and disappearing among swells. So do spruce logs with branches sticking up. The tour boat operators call those "spruce whales."

The ocean is breathless, almost asleep. On the hydrophone: surface noise, outboard whines, rocks roiling on a distant beach, a crackling sound that we jokingly call "crabs walking on the bottom."

On shore, a sea lion scratches himself with a hind flipper, a great, fat bull on a rock, as the gray swell touches in and turns white and breaks apart into fragments. In the book Craig's reading, Morrie tells a parable of a wave rising toward the shore in sudden terror, forgetting that it's not a separate wave but part of the ocean. We're like that too, out here, absorbed by what we want the ocean to give us. And I'm like that when I think ahead, to my new life, powerless as a glass bottle riding a breaker to a rocky shore.

Later, Eve comes out on deck with speckled blue tin plates and a pot of macaroni and cheese. We eat and talk to the sea lion, and about the sea lion, in strange, idiot voices. "He's fat, but he's nice." I tell Eve about the time Craig saw a porcupine and said, "He is very spiny. That is why he is all alone." Eve tells us that in her novel, a brother and sister begin to fall in love. Craig says, "When people get locked away together for too long, weird things happen."

That's what happened to us. We spent thirty days on *Whale 2,* just the two of us one summer, and we fell in love, and the shape of our lives dissolved and has not reformed.

I suppose that I'm grieving. I put down my tin plate and stare at the sea lion on his rock. For nine years, I knew my life's shape. Fall through spring, I lived in an octagonal house with my husband, John. Branches scratched the triple-paned windows. Because the foot-thick walls were paneled with tongue-in-groove spruce, it felt like we were living inside a huge tree. Once, at night, a boreal owl battered at a window with its wings, trying to get in. The house was in a birch and black spruce forest, and little paths led to neighbors' houses, and we'd travel them to have communal suppers, to help butcher moose, to bring soup or orange juice to someone with the flu, to watch a movie. In winter, the cold ritualized, shaped, framed our days: splitting firewood, stoking the stove, plugging in the car heaters, raising and lowering insulating window quilts. I worked on my master's degree and taught at the university. John biked to his job at the elementary school every day, even at fifty degrees below zero. In spring, I migrated south, to the ocean, the boat, Craig, and the whales. Two separate lives, each whole, connected by a brittle sinew, like dried seaweed.

When I separated from John, I moved just down the road to a little log cabin I found vacant one day when I was running. One damp autumn afternoon, as I carried the last cardboard carton out of the octagonal house, I hit my head on the car's roof rack as I bent to heave the box into the backseat. The box fell. Pens, pencils, paper scraps, bills, my birth certificate, tax records, phone messages, quotations, sticky notes, photographs spilled into the mud. Crying, I stuffed things back into the box. "This is your life now," I thought, meaningless bits of paper, scraps, trash. Now the scraps have followed me, two cabins later, down to Homer, where I will arrange them in drawers in another desk, or throw them away. In thirty-six years, I've changed dwellings nearly thirty times. I want it to be otherwise.

A burst of radio chatter interrupts the thought. Craig jumps up, looks at me. "What did they see milling?" He rushes into the boat cabin, grabs the radio microphone, and calls up a skipper. But it's a finback whale he's seen, off Granite Cape, a long ways away. The radio talk is constant in Resurrection Bay, a background noise I mostly ignore, except for certain words, aural search images, indicating someone's spotted whales. Because the tourists desire most to see whales, but also puffins, glaciers, eagles, sea lions, seals, porpoises, otters, mountain goats, black bears, the boat captains report to each other everything they see.

"There's a good eagle at Caines Head."

"A couple of goats in Porcupine, real good viewing."

"Saw a minke off Toe Point, but he's no good."

"I coughed up a couple of hair-balls off the ship lift." Hair-balls are sea otters. Harbor seals are beach maggots. They call whales "blubber."

"Anyone got any blubber reports today?"

"Anyone seen any blubber in the bay?"

Larger whales are "wide bodies." Ones that are elusive they call "Old Iron Lung" or "One-Breath Wonder." Killer whales are "big fins." They call one another "Magic" and "Baitball" and "Buddy," "Zippy," and "Captain Cormorant." If they can't find the big-ticket species, they swap jellyfish or shark sightings or weather updates, comparing how many tourists they've "lost" to seasickness.

On shore, the sea lion groans, and hermit thrushes twirl sound like jugglers, and a crow rasps, its voice dry and coarse, like the rocks piled at the mountain's base. Craig jokes that the sea lion sounds like a gluttonous man with indigestion, not ashamed to moan his trouble.

When we can't find whales, *Whale 2*'s size matters. The cabin is a square space with a corroded oil stove we cook on, a tiny table, and a little counter with a faucetless sink that drains out the cabin side. Sitting on the engine box, I can write at the table. Nautical charts are stashed there, against the back of the helm seat. The engine, hidden under the bunk, occupies the cabin with us, breathes like we do. It steals all the oxygen, so I installed a little window over the bunk, so we can all breathe. Eve sleeps on the bookshelf, a narrow perch above the bunk, with two windows looking out the stern. Like a corpse in a coffin, she sleeps on her back with her hands clasped on her chest. Her feet touch one wall, her head the other. She's five feet four.

I lie in the bottom bunk, reading, and think of Morrie's philosophy, that once he gave up and stayed in bed, he might as well be dead. My body and mind wallow in that stagnant realm, despised by sailors, known as the doldrums, partly because we eat all day, and randomly, except for breakfast. We begin with good intentions of yogurt and muesli, but in boredom, soon take out chocolate or rice cakes, cheese crackers or cookies, and never feel hungry. My body aches from not moving. A cloud has detached from those draped over Cheval Island and now floats in my head.

Eve lies in her bunk, reading, sleeping, most of the day spent up there. I wonder how to impose some small purpose, some discipline on this day, which passes beneath us as a series of swells. Morrie says that every night when we go to sleep we die, and every morning we're reborn. I want to be reborn in this very instant, but I don't know how—jump off the boat into the water, like the sea lions, plunging head first, or feet-first like a woman I saw once, whose entire, tear-shaped body was swallowed and then released?

I envy the tour boat captains, with their travel plans, their times of arrival and departure, their lists of animals to find, their lunches at noon. One such skipper, Captain Keith of the *Alaskan Explorer*, offers to bring us dinner, and Craig says yes. The boat's huge, and hundreds of people stare down at us from the glossy white bow and take pictures as a plastic bag is lowered on a boat hook. We open it to find foil pouches of fried chicken, little square plastic containers of barbeque sauce, and cookies wrapped in cellophane. We eat them because they are a gift, not because they're good, and we enjoy them and compliment them. Craig says they're not too greasy. Eve laughs and says that was embarrassing, but not too much. Sometimes a wave meets the nearby rocks with a sound like a whiplash, and sometimes with a sound like wind. Sea lions growl and roar and look like bloated logs, washed up from China.

At six, we give up searching, leave the radio on, and drop anchor in Pony Cove. We throw our black inflatable into the water, and I tell Eve that its name is "Pet Pig." "Pet Pig," she repeats, baffled. I've no real explanation, but I tell her about a child I knew who said he'd always wanted a pet puddle. Craig oars us in to shore—no beach, just a jumble of rocks. "This must be some kind of back eddy," he says, as the rubber hull swishes through floating masses of popweed, eelgrass, plastic bags, sticks, and feathers. Logs wobble at the sea edge like passengers on a subway train. We scramble onto shore, haul the inflatable above tide line, and stagger over the angular rocks, looking at trash. Jammed under the boulders are masses of nylon line, little foam buoys, and larger, round metal and plastic ones. If Craig's nine-year-old son were here, he'd want to bring them home "for the buoy man" in Homer, a collector of beach wrack.

We pick up flotsam, toss it aside. A plastic bottle from Japan. A gray plastic jug that Craig saves even though it's filled with foul water. Bits of wood and trash. The tide doesn't do its work here, of flushing. Nothing gets out of this eddy, no matter how useless it is.

But there's one treasure. As I jump from boulder to boulder, peering beneath them, kicking piles of kelp to see if anything's hidden inside, I spot a roundness of worn green glass, and think at first that it's the ultimate sea treasure, a Japanese fishing float. I reach in and pull up a bottle—round with a short narrow neck, dark green with pock marks from striking rocks, and a piece of glass shaved off the bottom—sealed but with a bit of liquid inside. Craig says it's a saki bottle. I say there's a spirit in the bottle who'll hear our wishes.

Up in the woods, Craig seeks space, hikes away from us and lies down. Eve and I clamber around moss-covered boulders. Thick tree roots, like uncoiling boa constrictors, clutch the rocks. We find one that makes a perfect bench and sit on it, side by side. Like a net of pale green lace, hemlock branches filter the light around us. Eve, the daughter of Craig's first marriage, spends every summer with him, but this year, she's not returning to her mom in Gustavus in the fall. She's staying in Homer with Craig to finish high school. I tell her why I've been sad and quiet on the boat sometimes, how I'm scared of moving to Homer, of not knowing anyone. "I can understand that. I'm scared too," she tells me. "I'm going to miss my mom and all of my friends."

"And what about Shreve?" I ask, half teasing, but Eve's serious.

"I wish it could be like when I was Elli's age, when dating meant holding hands, talking on the phone." Elli, Eve's sister, is twelve. Big tears roll down Eve's face, and she suddenly looks like she did when I first met her, five years old, her round, pink cheeks and large brown eyes peering from beneath the hood of Craig's stiff oilskin jacket, which drug on the ground around her, the sleeves folded in half, her hands still not poking through. Swaddled so, she stood on the dock in pouring rain while we replaced an engine.

As she rests her head on my shoulder, crying a little, the bay suggests itself as glints between branches, a few crows shred the air. It's like we're folded into a pocket of time, into a killer whale's stomach pouch, holding all indigestible, stubborn things: bones, whiskers, salmon scales, porpoise skin, the things we don't want to give up, the things we want to be different without going through the process of change itself.

After I left my marriage, I dreamed of fires and snakes. Once, I dreamed that I watched a woman, holding her child, jump off a burning ship at night. In other dreams, snakes chased me in and out of gardens and swamps, and I thrashed away. My therapist, a Jungian psychoanalyst, said that fire and snakes were symbols of transformation, that I needed to stand in the fire, to let the snake strike and break my skin. Transformation is a long process, she told me, but I wanted to wake up one morning, green and tender, free of my old hide.

Something black moves in the forest. "Eve, look." She turns.

It's a bear, lumbering through the blueberry undergrowth. It pauses to peer up a steep brushy slope, as if considering, then arm over arm, flattened against the green, glides up the slope like a cloud shadow. Craig's sitting up, looking too, and then it's time to go.

That night, after dinner, we sit in the cabin. A tour captain relays a message over the radio, a report of a large group of killer whales out in the Gulf, feeding like crazy on enormous salmon schools, twenty-five miles offshore. Craig switches the radio off.

Light rain falls outside. I'm crouched on the step, holding my empty plate, my back against the wooden door. Water seeps in at the bottom seam.

Craig's sitting at the helm, his long arms and legs spidered over the wheel and cabin entrance, and Eve's perched in front of the bunk. Our dirty dishes cover the little counter, and steam from the kettle fogs the windows. I pry open the lid on the saki bottle and sniff. It smells like holy water or tears. We take turns holding the bottle in two hands. *Make one wish for yourself, one for someone else, and one wish for the planet. Between each wish, blow three times into the bottle.*

I think of what I read today. Gaston Bachelard said, "What special depth there is in a child's daydream! And how happy the child who really possesses his moments of solitude. It is a good thing, it is even salutary, for a child to have periods of boredom, for him to learn to know the dialectics of exaggerated play and causeless, pure boredom." Craig's round eyes widen in earnest surprise as he blows into the bottle, and Eve's turn upward, toward heaven as she holds the bottle to her pursed lips like a chalice.

"Let's put on our lounging pants and bed socks," she says, "and get all cozy and read!"

Morning, lying in bed. "Tell me stories from your childhood," says Eve. So I tell her, again, how I once filled a claw-foot bathtub with wormy apples. Old Mr. Robinson used the bathtub to water his three horses, two ponies, and an incessantly braying donkey named Kernel, whose long ears poked through his straw hat. I thought the horses would like the apples, but they made them sick. They were green. Mr. Robinson was very old by then, and infirm. The horses' caretaker, a cruel woman in tight jeans named Betty Frey, called the police. Eve laughs from the bunk. "She really was cruel," I tell her. "She used to whip the ponies, and once, during a heat wave, she locked the horses in the windowless barn for days. My sister and I crawled under the big door with knives, sliced apart the baled hay, cut the ropes holding the swinging doors shut, unlatched the stalls, and set the horses free. Betty called the police again."

Craig tells of Eve's grandmother Doris, a passionate environmentalist and fastidious housekeeper who "always" wears cleaning gloves. He remembers her wiping the ceiling vents, then standing on a chair and drying her hair beneath them to save energy. Craig and Eve's mom would scrub their hands together vigorously, say, "Doris is going to clean up the earth, scrub it until it's clean." He told of his Irish grandmother, Connie, who refurbished Victorian houses with her carpenter friend, Mr. Wurr, a stooped old man who never spoke, just did her bidding. Connie took Craig to drive-ins for burgers and hot fudge sundaes for breakfast. I tell them about my grandmother who died at 104, a peasant woman from Latvia who lived in the United States and Canada for forty years but never learned English, who filled gallon coffee cans with apple slices she'd dried on cookie sheets in a warm oven, the same wormy apples that sickened the horses, picked from the ground beneath Mr. Robinson's trees.

"I like when we tell stories," Eve says. "Sometimes the stories are so much better than the things actually happening, than living through them."

"I like stories about whales. But what I like even better are real whales," Craig says, starting the engine, turning on the radio. "Where are the whales? I need whales. I'm going craaaaazy."

As we leave the bay, I watch a puffin try to land on a cliff face. There's no purchase, so it pushes back off from the rock in midflight, its red feet like a swimmer's making a kick turn, and buzzes off, flying with the sheer daring of stunt pilots, wings whirring like airplane propellers.

Hours later, after lunch, the sea blank, refusing to yield up fins or blows, another imperative. We need exercise. We point the boat's bow across Resurrection Bay, toward Rugged Island. All around the bay on cliff tops are World War II gun emplacements, Craig tells us, and soldiers were stationed there for thirteen months at a time, in bunkers made of oozing concrete with iron doors and narrow window lookouts, waiting for the Japanese invasion that never came. I imagine them staring through scopes at the horizon every day, eating up their rations, playing cards, growing dirty and bearded, constantly wet and cold and bored. At the south end of Rugged Island, in Mary's Bay, where we're headed, an old trail zigzags up the mountainside. We can hike there, to the bunker, Craig says.

And now, on our way in to shore to try to find those bunkers, the sky lowers, so we're in the clouds. Everything blurs, even the green, even the cormorants' edges, and the white on their rumps, softened, like our minds, or like brie, which Craig calls "fat boy cheese." Since it's calm, we tie up to a decrepit dock built during the war to service the bunkers. On the tilted pilings, which still ooze tar, a miniature forest grows, and as Eve points out, even a garden of moss and flowers, like a pin cushion. We row beneath the dock, and I suggest we have a picnic up there, even though jagged, ripped wood clearly marks the rotten places where we'd fall through. I imagine my wicker picnic basket askew and bobbing on the waves, the three of us vanished.

From the water, we can barely see the shape of the old road leading up the mountaintop, switchbacking in multiple Zs of paler alder and salmonberry—the plants that colonize what's been bulldozed and left behind.

"It's going to be a nightmare bushwhacking through that," Craig says, but the trail, we find, is maintained, branches cleared away, easy walking. My legs hurt deliciously after three days on the boat, the muscles stretching and unwinding. I take deep breaths, feeling like a coal miner emerged from the pit.

"Who maintains the trail?" I ask. "What if one of those army guys is up here? What if he got left behind, and he's still waiting for the Japanese? If you listen, at night, you'll hear his voice keening, *the Japs are coming . . . the Japs are coming.*" I describe him creeping, grease-blackened face, among the grass and salmonberries, hiding out in a bunker. Or maybe he's a rattling skeleton, jiggling his bones as he scampers through the underbrush.

"It's funny how you think, Eva, always that there might be something wrong," Eve says, stopping to pick salmonberries.

At the ridgetop, we find a subterranean cement shelter of three rooms built under the ground accessed by a square chimney-like opening at our feet. Seen from below, it juts out, overhanging a thousand-foot cliff. Eve goes in first, hand over hand down rusted grips bolted into the gray concrete. Her teenage aesthetic is that she loves it down there. She likes the graffiti and finds a rose and many names and dates, like Zena 96, Winnie the Pooh, Boona and David, True Love Forever, and we wonder who are all of these people? How did so many teenagers get here to scrawl their messages? We're twenty miles from town by boat.

In my eyes, this empty place is a weird, accidental aerie. Gulls glide circles, and a pair of peregrine falcons sweeps a hundred feet above the bunkers, teaching their two chicks to dive at the gulls, and a sea lion roars a thousand feet down on the ocean side, and sea foam swirls green and white, the very same water swirls and swirls there, shoving and nudging and licking at the rocks, and sometimes it seems that the sea pours out of a cave, pours and pours and is the source of the sea. We peer at this through a narrow rectangular opening in the bunker. Where the guns once perched, where the soldiers stared, long-winged birds slice at the view. The floor is concrete and clean, as if someone had brushed the detritus—leaves, twigs, cement rubble, rags, an old shoe—into piles in corners and along the walls, and we wonder where the stove was, and think it must have been cold, very cold and lonesome here.

From this vantage we can see many miles, and we scan the Gulf to the place where the sky streams down to the sea in a squall. There is no horizon, just an uncertainty where clouds and water blend, where the Chiswell Islands hover like scattered ellipses off the end of Cape Aialik.

On the way down, I walk behind Craig and Eve, stopping to watch them round a switchback until they're out of sight. Even their voices bend the corner, like birds, and disappear, so I can only hear the breeze through the hemlock and spruce needles, and below, the surge of swells sucking the shoreline of Mary's Bay, and the thud of *Whale 2* against the dock pilings, and a gull swarm shrieking over a bait ball. I'm alone. The details of my life slip away. I know with certainty, someday I'll be truly alone like this. Craig is eleven years older than I am, and if we're lucky enough to be together until one of us dies, I will most likely be the one who's left behind. I make myself old. I let the aloneness settle in me like cold water filling a crevice at rising tide. In the book Craig's reading, Morrie knows he's going to die. There's no way he can get out of it. He knows his body is going to deteriorate until someone has to do everything for him. He keeps on going, even in his chair, into each moment, into the minuteness of each moment passing, even in pain. I think of those tufted puffins, the way they toppled headfirst from their rock ledge, pointed like arrows with two white stripes toward the water in an unswerving path, and the three guillemots, who braced themselves and arched and dove straight down.

Suddenly, the future and now angle through each other and me. Craig is gone. Eve's moved away. I'm an old woman. Am I content? I break and run down the trail to catch up to Craig and Eve.

When I find them, they're bent over the inflatable's line. The perfect, fail-proof bowline we tied worked free somehow, and the inflatable, completely grounded before we left, floats, bumping the rocks gently. Luckily, it hasn't drifted away. "It's the old army guy," I say, "the one who maintains the trail. He untied the bowline so that our raft could float free, so we'd stay."

"Oh brother," says Eve.

In an hour, we're cruising again, scanning, listening, thinking maybe now the whales are here. Jostling in a tide rip, hydrophone down, a quarter mile offshore, we wait. Suddenly, a bumblebee zips by, circles the boat, perches briefly on the gunwale, fanning its wings, then launches itself back toward shore.

When nothing happens, we try to make things happen. Not just here, but in the "real" world—jobs and dramas, breakups and breakdowns, relocations, reunions, departures, mountain climbs. We drop structures—nets, purposes—over our lives. Whenever I leave this place, I watch closely the animals who stay, and I long for their fidelity. Bachelard says, "At times when we believe we are studying something, we are only being receptive to a kind of daydreaming." I tell Craig, to reassure him, that zero data points are data points nonetheless. We still fill out our logs every night, sketching our search area on a map. *No whales seen. No reports of whales. Whales reported out in the Gulf.*

All day, I type into the computer, not the data points of our desires, but those of the actual, minute, and random experiences of our lives, which are half witnessed and meticulously observed, and half daydream. A friend of mine says our whole lives are dreams. At night we dream in one sea, by day in another, and which is real? What continent separates one kind of dreaming from another, ourselves from the places and the past we think we leave behind?

Shapeless day, filled out, ballooned, like a jib. What fills it? Puffins perched on ledges, dive-bombing from their roosts. Cormorants, wings outstretched like cloaks of dunked witches. Thoughts enough to fill submarine canyons, passing meaninglessly, as significant as each individual wave, but, one thought, built on itself, big enough, in the moment, to coax death.

Two days later, our last night on the boat, we anchor in Porcupine Cove. No one's seen any whales, near or far. Although this place is open to the Gulf, and swells break whitely on shore, Craig guesses we can land the inflatable, so we dump it overboard and climb in with our paddles. Closer in, we see that the beach is steep. I watch a wave curl up, sizzling. As the next wave builds before striking land, it sucks the last wave off the beach. Craig rows. We wait past the big sets.

"I don't know about this," I say, and Eve looks around nervously, one hand on each pontoon.

"No problem," Craig says, his head craned back, oars poised, and then he strokes hard. There's tension as we rise with an incoming wave, pausing on top, and then careen toward shore, grounding as water and stones pour back down the slope. We scramble out quickly before the next wave comes. I laugh at Craig's jeans pulled up so they look like capri pants, his shins white and hairless where he burned them playing with matches and gasoline as a boy, his huge bare feet in sandals.

Behind the steep beach we find a lake. Scraps of sea paper—sheets of algae dried into chalk-white mats, imbedded with leaves and grass—litter the lake edge. I imagine our thoughts scratched onto such thick tablets with soft pencils, like those I learned to write with. We lie propped on our elbows, tossing stones. We say we should swim. It will feel so good. We dip our toes into the water and draw them back and say it's cold. But not too bad. It's really not that cold. Well, pretty cold. It would feel good though. Okay. We sprawl on the stones and watch small fish. I call them guppies, because their bellies are fat and hang like sacks. Craig calls them sticklebacks. Eve watches a group of them fight over a fly. And then it begins to sprinkle.

There's a log on the edge, a huge log standing on its side, on stubs of what were once branches, like a trivet. And we shout to hear echoes against the rock wall, and the sound of the surf reflects off the wall like a reply, so it seems as though the sea is coming in, announcing itself, and all those logs are evidence that the sea, in winter, does come in, rises up and over the steep beach, throwing logs down onto the lake, where they float, meander, and nudge the shore like stranded whales. One log balances on a boulder, and as the water laps in and out, it seesaws on its fulcrum.

We're all poised here. We want to take off our clothes and plunge into the cold water, but inertia holds us back, and then it rains and we're excused. I want to be settled. I want to know my purpose in every moment. We're poised, waiting for the field season to begin, waiting for our lives to begin. We wait for healing. Wait for courage. For our blundering and mistakes to be done, to be our best selves, at last. Eve's not ready to grow up. We want time to stop, or at least to invert itself, to transform us. The moment is so full that we get up and walk away, pulled by the desire for bed socks and pajamas and hot tea and books, for the comfort of that.

On the boat, I sit in front of the glowing laptop screen, rereading my words. I realize that I am writing this for Eve. And what I'm trying to say is that during the times when nothing is happening, we're like the saki bottle with its tears bottled up inside. In transition between one life change and another, we're like field mice who see only what's right in front of their beady little eyes. During my first summer of whale research, when I was twenty-three, I used to collect rocks and shells, carry them in my pocket. On solitary hikes, I'd find a spot to kneel on the ground and make a stone circle, inspired by a book about medicine wheels. Around the circle, I'd place, in the south, a piece of green foliage, in the north, a white shell, in the west, a black pebble, in the east, a yellow flower or seedpod. The animal of the east was the eagle, who soared above and saw with perspective, who saw the big picture. The animal of the south, the place of innocence, was the field mouse with its gift of detailed examination and curse of no perspective. When I want time to turn and flow and fill the bay again, I have to be like the field mouse, concentrating only on this moment. Sometimes I have no choice. It's all I can do.

I read what I've written out loud to Craig and Eve. "What will the end be?" I ask. Eve and Craig say that the end will be when the whales come. I say the end will be when we go into town, whatever happens, even if nothing. But if we believe Jane Hirshfield, that actual experience has no firm resolutions, no flares of instant transformation, no spontaneous human combustion, then there is no end. An end, in an essay, is essential. An end in our lives—that is, until we die—is make-believe. We can't ask Morrie anymore, because he died, though in the little book of his words we still contemplate him contemplating the last time he'll be able to walk by himself, crap by himself, feed himself, recognize his loves and hates, wonder what comes next.

Near our anchoring spot that night, seals sleep on a nearby rock, their bodies chalked against the gray boulders, their faces with big dark sockets like skulls. They look like the skeletons of sunken sailors, drowned at sea and risen. All night long, they groan and snore, and words and a name repeat in my mind: at night we sleep with the seals, at night we sleep with the seals, at night we sleep . . . every night we die, and every morning we are reborn . . . are *resurrected*. Bachelard exclaims, "How often have I wished for the attic of my boredom when the complications of life made me lose the very germ of all freedom!"

And that's why they call this Resurrection Bay, a place where, when nothing happens, a moment garrets us in absolute boredom, expands to brim its meniscus, until we can't stand it. Resurrection Bay, where, in boredom we see most minutely, where we live most minutely, as minutely as bees circling the boat at sea, as pebbles roiled round and round, smoothed until our purposes are gone.

"I'm going to call this essay 'And suddenly, nothing happened,'" I tell Eve.

"It seems like nothing's happening, when so much is happening," she says.

Seven Januaries

1 January 2004, Kohala, Hawaii

> *Another year gone—*
> *hat in my hand,*
> *sandals on my feet.*
>
> —Basho

The ocean feels restless today, unsure of what's to come. There's a lull between weather fronts—or maybe the end of a pattern—we don't know. The ocean's a murky blue-gray slurry of phosphorescent turquoise edges and white frills. I'm lying on a bed of ironwood needles, a few feet from the cliff edge. My feet are dirty from yesterday's walk, a rust-colored, permanent stain. Above me, bleached arms of salt-bitten ironwoods intersperse the pale green ones, battered by winds and spray. To the south, swales and hills undulate above russet-pink cliffs, their colors muted, silvery. The sun breaks out, changing everything.

Last night, on paper scraps, the kids and I wrote what we wanted to discard from our lives, then came here at dusk, tied the scraps to dead wood, heaved it over the cliff edge, into the surge below. And toward the void opened by those leavings, something is now rushing.

Today I am alone here, barefoot, arms around knees, looking out toward Maui, waiting.

13 January 1998, Iowa City

Earlier it had been bigger, low, pale. As Molly Lou and I walked, in the evening, it distended as it sailed behind branches.

In the huge, white hours of night, it wakes me, staring in from the corridor of sky between Molly Lou's apartment and the house next door, painting squares of light on my arms. I get up and look across the alleyway into windows of the neighbors' apartments. In one, a cluttered office, a man sits unmoving at his computer under a blue-white fluorescent. In the lighted window of another apartment, I detect only the edge of a staircase, which no one, at this moment, is ascending or descending. I clutch my elbows, nightgown flapping around my legs from the warm air of the heat duct, consider divided habitations, each walled and lit, strangers awake in the middle of this night. Let's all walk outside, meet on the sidewalk, in our pajamas.

I crawl back in bed next to Molly Lou, who's still sleeping. I lie a long time thinking about my mother, into whose life morning will come two hours sooner. Time zones divide us. And habits. And walls.

When does a year begin?

In the morning, the sky is pale. The sun is warm enough that I smile. Just a crunch of snow on the ground. We all walk under a larger sky. It is the same sky.

1 January 2004, Kohala, Hawaii

A snowy morning—
by myself,
chewing on dried salmon.

—Basho

I stay here some days to write when everyone goes to the surf beaches. During my long walks, pangs of loneliness, like wind-puffs that flutter the prayer flags, startle me. Sometimes I *know* that I'm in the middle of the Pacific, that this island's a ship, and it's drifting toward a sky in the distance. I should ride it.

But after a few hours, restlessness pricks at me. I walk to town to mail letters.

6 January 1999, Chiloe, Northern Patagonia

While I am staying with my friend in a cabin on the island of Chiloe, my mother has a brain aneurysm. But I don't know about it. At night, my mother's monitors blink and beep. On Chiloe, my friend and I light candles, sweep spiders, tell stories to the two young boys staying with us. They panic when we go outside to pee. There are spirits, they say, these good Catholic boys from the city. They describe

Trauco, who makes girls pregnant,
Caluche, the ship phantom,
Colocolo, who won't leave a house until everyone is killed,
Pincoya, the sea goddess,
Viuda negra, the black widow who hangs her web in lonely places,
Brujo, the man-witch, who cast spells and lives in caves.

Back in the city, having heard the news, before boarding the plane, I kneel at the Virgin Mary statue in the Concepción cathedral, praying for my mother. All around me, others are praying in the dark church. One man falls to his knees. *Colocolo*, leave this place. Leave all of us.

But *Colocolo* stays, sweeping out my childhood house, for years. He says this is only the beginning of winter.

3 January 2004, Kohala, Hawaii

Reality is these damp feet in socks tucked under my thigh and shin. Soaked towels draped over branches of yellow-flowering bushes in the rain. The rust stains leach up my ankles. I soap them but the streaks return, like ghosts. In a poem, an Inuit daughter describes her mother skinning a seal every day on a little table, scraping fat off the skin with her ulu, putting a piece of fat on her daughter's tongue. Imagine this mother's hand, softened by seal oil, impregnated with the sweet-rot smell of marine mammal, the table and the mother's wrists slick with oil. Think of my parents, reeking of pig fat for days after a butchering, my father's hand smelling of smokehouse, of bacon. At some point, a person has to give up the constant scrubbing. The rusty bare feet walk back into our tent house every day. This is how it is on earth.

Back at home, mounds of perfect snow finally fell in December, then it rained and turned the roads to ice, pocked the snow, ruined its perfection, made muck in the chicken coop. I lie back and listen to the rain get harder, die back, get harder again.

Despite intentions, the old habits.

Resolution: tomorrow, I'll do it, write deep into (not around, not past) my boredom.

New Year's Day—
everything is in blossom!
I feel about average.

—Issa

20 January 2000, along Turnagain Arm, Alaska

We glide through eddies of fog on the drive to Homer, the past enveloped in layers so that the mountains are fading, a sense of things never resolving themselves before they start turning into other things. Flexible, in flux, indentations in the sky become a row of knuckles, crenulations, the sky folded, refolded into a flank of mountains, then into whiteness. I gasp, startling Craig. He brakes. "What's wrong?" Then sees it. A mountain slices the full moon, a solid bronze disc, in half, then swallows it. Darkness gapes and swallows us. Then, out of the fog, suddenly, the moon is back, resurrected, and there's clarity, but we know it's an illusion. The place of clouds and change was more urgent, more real. Here all is angular in harsh light, in black and white, and we think we know things. A mountain will be a mountain. As the road curves, more of it will emerge. The moon will be the moon. It will not dissolve or resolve into a complex gesture, a hesitation, a suggestion of something. There, everything was fluid. Now, looking closely, the mountain range looks like a rack of satin wedding gowns, a procession of brides. Now they will marry the sky, like the woman who paints her face white and returns to a state of virginity.

This is what deep winter does. It takes our minds. Unskins us. This is its purpose.

14 January 2004, Homer, Alaska

Zero degrees. It settles in, winter in its profound sleep. At 11:00 a.m., smoky gold light bulges into the kitchen. At night, cold presses against the walls. The sun still carves a low arc over the mountains each day, backlights spindrift billowing off the peaks. The mariner's forecast warns of heavy freezing spray, gale-force winds. The local radio announcer, mining for news, interviews physical therapists about female incontinence for far too long.

In *Soul Mountain*, the narrator, diagnosed with fatal cancer, then mysteriously cleared of the diagnosis, undertakes a 15,000-kilometer trek through remote regions of China asking, *Who is this self?* "The existence of an other resolves the problem of loneliness but brings with it anxieties for the individual . . ." the translator suggests. I am engaged with this power struggle, between the material world—my stepchildren, my job, housekeeping—and my writing, which requires loneliness, contemplativeness, isolation. When writing, such distractions anger me. Then I get lonely. Should I have a child of my own? When engaged with my family, sometimes their demands anger me, and I shut my door. But when my stepdaughter worries over her high school finals, compares herself to classmates who know more, who memorize quickly, for whom it all comes much more easily, I tell her, "For you, kindness comes much more easily. And that's what matters."

The Buddha said the myriad of phenomena are vanities. And the absence of phenomena is vanity also.

14 January 2001, Christmas Tree Burning, Sitka, Alaska

Among strangers at the bonfire pit, I chat about writing, about kids' school projects, try to insert myself into conversations between friends, to whom I am unknown. My gut's in a knot, hating this kind of gathering, knowing only the hosts. I tell someone I liked his poem, used it as an example in my class. Someone knew my ex-husband when he lived in the bush. But how do you really get to know people? A few of us stay at the fire pit longest. Staring at the flames, poking at them, a distraction that makes it seem not so awkward to have nothing to say. The green needles spark and crackle, constellate against the dark, drift off. I toss in old essay drafts, a copy of my master's thesis, page by page. A man says, "You're burning time."

3 January 2004, Kohala, Hawaii

Fourth morning of rain. Kids sleep on a big, communal bed made from cushions shoved together, the smallest girl fallen into a crack, the tallest, her face hidden by waterfalls of hair. Half-waking to the sound of rain on the tent roof, they close their eyes and minds again and drift as though they're at sea on a lifeboat, and there's nothing to do but wait for rescue. Even in mildewed sheets, they find warm places, in dreams, in the heat of each other's bodies, radiating their small fires to dry the damp places between them and at the napes of their necks.

Steamy, seventy degrees, geckos clacking at night, luffing Kona winds—they call this winter here.

1 January 2002, Arroyo Seco, New Mexico

Three kinds of selves, of light, the way what we see is just surface luminosity, an aura, or light passing on from one place to another—ourselves (passing through) on our way to somewhere else. The sky is so big here, oceanic. It dominates the paintings too. The sun's burning through these high clouds, illuminating the table. When it warms, the little adobe house creaks, as if there's a cougar on the roof. Life passes along the palette of sky, scrapes away at the earth, slides between branches, light in tides washed over trees. Everything lightens against the land as though chalk-dusted.

Why am I alive? I'd forgotten the question. Forgotten the standard by which I measure the quality of my life—a painting?—a state of mind? The way the light and sky change very slowly, and I with it. When it walks with me. Must everything grow bigger? We believe too much in a certain kind of evolution. Now amazing winds and anvils take shape in the sky above the mountains. Can't we hold still awhile? Tolstoy said, *True life is lived when tiny changes occur.*

2 January 2004, North Kona, Hawaii

Luminous light on the water. Nearly five in the evening, we recline on dingy beach loungers in warm drizzle while Craig swims and the sun goes down. The light is violet-green but changing. Waves curl up green, spill over brown. The kids read, I scribble, but we're all distracted by the gray-haired man jogging in trunks with tiny steps, back arched, head keeled back, belly thrust out. Also running back and forth on the beach, a middle-aged woman in a skirt and loose maroon blouse clutching a fruity drink—plastic cup half full of pink, straw dangling—in one hand and a purse in the other, as though trying to hail a taxi. As the sky purples, workers light torches at the resort, and the old Chinese lifeguard saunters toward us, collecting his flags, pauses behind my chair, taps me on the shoulder, grins, says: "As he walked by, he turned to smile . . . a very phony smile, but still, a smile. A nice lifeguard guy." Touches my shoulder again before striding off, brown, buff torso in orange shorts visible a long way off.

Boogie-boarders wait for good waves in the Easter egg sea. Then it turns even more purple, baby blue between each wave. Over it, a dark cloud aslant. When big waves crash upon the swimmers' heads, the ionized air drives the salt smell in our faces. I breathe it deep. Rain sprinkles on us, sheens our skin, but we don't care. When Craig comes out dripping after fifteen minutes, we send him back in. Gradually, the colors leach out of the ocean, the light grows dull. The three of us, content. Give us this day, again and again and again.

2 January 2002, Arroyo Seco, New Mexico

Here,
I'm here—
the snow falling.

—Issa

Three feet of snow fell in the night. We're trapped. To do list (in retrospect): Shovel, chat with man in purple jacket and black cowboy hat on front-end loader, eat pomegranate, write adventure stories of swamps and treasure maps, sweep wood chips, stoke fire, play cards, drink green tea, trudge to town in snow pants, pee in pricker bushes, click tongue at horses who look up, startled, make snowman with tumbleweed hair, light palm wax candles at night, wear two pairs of socks and pajamas all day, drink more tea, eat more pomegranate, this time, really fast with serious look on face, have red cheeks but don't know why, show off cowlick, snuggle in sleeping bag, dream you're a queen, eat sage honey, stare down bison, chew an entire bag of bubblegum, watch who's walking down the road with yellow dog (little girl in snowsuit), listen to wood crackle, worry that this will be a boring vacation for the kids, peruse the books on this stranger's shelf, read Gary Snyder's

Earth Verse

Wide enough to keep you looking

Open enough to keep you moving

Dry enough to keep you honest

Prickly enough to make you tough

Green enough to go on living

Old enough to give you dreams

Don't believe in resolutions, but make them anyway.

8 January 2003, Lahaina, Maui

Mother I never knew,
every time I see the ocean,
every time—

—Issa

Sitting on the park lawn watching him swim as the sun sets and a man walks by with two Dalmatians. In the ice plants, a green caterpillar pulls itself along stems, chewing. A mouse suddenly attacks it, quickly retreats. A humpback floats offshore. The sun descends toward Lanai. On my left, an old couple, the woman knitting, the man sketching. On my right, the constant murmur of a dreadlocked man talking to someone I can't see. Who I can see is a woman in a wheelchair on the outward edge of his conversation who makes me think of Mom. She looks to sea. From her clawed left hand, the smoothness of her facial skin, I know she's had a stroke. An older woman and two men in Hawaiian-wear step onto the sand, the men in docksiders, Aloha shirts, leisure pants, talking animatedly, arms waving. Another old couple in lawn chairs sits reading. From the resort development next door, a pile-driver clangs. The moored sailboats sway. A couple holds each other, watching the sunset—now they kiss. For a moment, I am alone on the earth among strangers. No one knows my history or has any thought of me at all, not one expectation.

There could be physical walls between us, and it wouldn't change one thing.

The sun contracts to a streak like clarified butter spilling toward shore. The natty trio comes back, emptying sand from their shoes. The woman in the wheelchair points to something, but no one's paying attention. I look out. His arms make gold splashes on the dark sea. Looking again at her, I miss my mother terribly. Now the sky above Lanai turns the orange of coals. Down goes the sun. Lanai turns blue. Everything is dimming.

11 January 2004, Homer, Alaska

New Year's morning:
the ducks on the pond
quack and quack.

—Issa

What awakens the mind first? Maybe the still-warm egg in my pocket. Maybe the pewter light an hour before sunrise. Maybe the sound of the 9:00 a.m. Twin Otter buzzing to Anchorage, like a saw rasping a big, serrated cut into the air, air still silted with the ink of darkness. Maybe the dark houses of the neighbors, sleeping in on Sunday morning.

And then the distractions, the hooks I swallow, that yank me back up into the material world. Crumbs on the kitchen table. The glaring, artificial sunlight-box that makes me squint, that makes my eyeballs ache (fake substitute for Kohala sun). Our little dog nosing in for attention, bobtail wagging its whole fuzzy back end. When I close my eyes, what's only in my brain: parallel stripes of blue-white light. January. Bleary world, a just-before-waking dream. A young moose licks branches outside, moving a few steps every minute, but through the long night, its wanderings punch a meandering drunkard's path over acres of empty fields. I ski along those tracks. While my body moves forward, gliding, pushing, tromping, my mind moves back, to all of those forty previous Januaries compressed, a book between tightly bound covers, calendar pages from every year. And the Januaries still buried. Burning times. Dark times. The coldest month. When everything pauses in stillness, in solitude, in isolation, preparing to enter the world again.

The old calendar
fills me with gratitude
like a sutra.

—Buson

14 January 2004, Homer

Intentions thrown out or burned. At my kitchen table. Noon. The north wind. I wait.

Epilogue: Letters to Mike

This is a cliff edge. This is a knife.
This blade, this little space: your question, questions.

—Molly Lou Freeman

There's a map in my head. In its center, a compass rose—a constellation of seven tiny islands, the Pleiades—marks the place where twenty-mile-long Knight Island Passage bends to the east. Every few seconds in the dark, the navigational light blinks on the northernmost island, showing me the way home.

I return to that place every summer to find the killer whales. I imagine the whales have a map in their heads, too, because they return to those same places, where the silver salmon migrate, where harbor seals haul out or have pups.

Whale Camp once was used by people whose inner maps included place names like *Egagutak, Iktua, Tyaigyulik*, who called killer whales *arllut*. Their ancestors lived on Chenega Island for thousands of years. When I asked people from Chenega about the place killer whales have had in their culture, several said, *They have always been here.*

It's as simple as that. Or is it? From the beginning of the oral history of the Sugpiaq people—a group inhabiting Prince William Sound, Cook Inlet, and Kodiak Island—killer whales have been there, in experience, in art, in story, in knowledge, in belief. The same is true for the other coastal Native cultures of Alaska: the Yu'piq, Eyak, Aleut, Inupiaq, Tlingit, Haida, Tsimshian. As long as humans have lived on this coast, killer whales have been here, both species' lives dependent on the sea. Sitting in their skin kayaks, hunters, in silhouette, even resembled killer whales, wore wooden hats carved with killer whale symbols.

They have always been here. But since I began studying killer whales in 1987, things have changed, and it's no longer clear that generations from now, people will still be able to say that. The oil spill irrevocably altered the ecology of the Sound and the psychology of its human inhabitants. The number of resident killer whales is growing, but the AT1 transients are disappearing. Their blubber is toxic with PCBs and DDTs. Other transient killer whales, genetically distinct from the AT1s, sea lion and whale hunters who range thousands of miles through the Gulf of Alaska, are also in decline.

Scientific research has lead to many new findings, and our knowledge has grown. Our appreciation for the complexity of the species increases with every new discovery. But all of this knowledge can't stop the extinction of the AT1s. Despite our knowledge, oil tankers still carry crude oil through the Sound. Despite what we know, the U.S. government refuses to sign a treaty banning the use of PCBs and DDTs on a global level. Despite tables and graphs and long-term data, the ocean itself is warming. What is the purpose of knowledge, of science? I struggle with disillusionment. Every summer, I return to the Sound as a field biologist, but a sense that something is missing has grown in me, so I've turned back to the first people to live on this coast, looking for help, looking for a new way to look, to understand.

When I was a graduate student, I returned to Whale Camp one evening, and a friend who'd stayed behind told me a visitor had stopped by, a man from Chenega with a dead seal in his skiff. He knew I was studying killer whales. It was Mike Eleshansky. Over the years, I got to know Mike, and when he died prematurely of cancer, I was bereft. I began writing him letters as a way to ask him things I never got around to, to help focus my mind on my questions about science.

I'm looking at a black-and-white photograph over my desk, of Mike's birthplace, Old Chenega, some time before the tidal wave destroyed it. His ancestors began living on Chenega Island at the end of the last ice age, according to the late Chief Makari. When the retreating ice exposed the small islands called Kalugat, perhaps the Pleiades, a chief went to look for a new village site in the area, and Chenega was established.

The photograph depicts Chenega many thousand of years later. The ice age glaciers had retreated to the heads of fjords in Icy Bay: the Chenega Glacier, the Tiger Tail, still releasing bergs into the Sound, a reminder of those earlier times. The photo is from the last decades of Old Chenega, sometime after electricity. Alongside dead trees, spindly power poles tilt, concatenated down the slope, dissecting plank houses with corrugated roofs and sheet metal chimneys. No smoke rises from them. It's spring. Dead grass tufts poke through decaying snow. Two children, a girl and boy in hats and coats, walk down a snowy boardwalk, their backs to the camera. The boy has paused to kick at the snow with his foot. The girl watches. She looks like a tiny Zen nun in a black robe. They might be descending from the school above the village, the only building spared by the quake. Mike was in his twenties when the tidal wave hit, when the eighty-foot wave surged, sucked back, came again, wiping out the life he'd known.

The glacier, whose retreat invited the Chugachmiut people to live on Chenega Island, by its monstrous wave, turned them away. The killing tidal waves came from Icy Bay, where seals still go to molt and have their pups on broken-off bergs, where Mike used to hunt, the years I knew him, and where the last AT1 transients, singles or pairs, survivors of unnatural disasters, still hunt.

After writing this, I read it to Craig, who began studying killer whales in Prince William Sound in 1980. I told him about the settlement of Chenega following the ice retreat, and he said, "Hmmmm, that's probably when the killer whales came back, too, the residents from British Columbia and Puget Sound." He related recent evidence showing that present-day resident killer whales in Prince William Sound share genetic haplotypes with British Columbia and Puget Sound residents. The AT1 transients, on the other hand, probably came from the Aleutians and farther west. Transients with that haplotype were found in the Bering Sea and Russia. And for humans too, the Sound was a meeting place for cultures: Aleuts from the west, Tlingits from the southeast, Eyaks from the north, coming down the Copper River to Cordova, and, from the southwest, the Sugpiaq people.

We exchange stories, Craig and I, in a way that enriches our understanding of the relationship between killer whales and the Sound, what, as an undergraduate, I learned to call ecology. In a three-way exchange, what would Mike tell us? Into this killer whale ecology, what dimension could he add?

Dear Mike,

Last night I dreamed that I went to hear a Native man speak. What he said was so true, so compelling, that, when he finished, everyone in the audience clamored after him, and all he wanted to do was escape. I pushed my way forward to ask him what he knew about killer whales, but I was too late. He was on the back of a moving pick-up, waving.

That's how I feel about your knowledge. Am I one of those clamorers, white folks, some wearing Inupiaq parkas, some in southwestern regalia, fingers hobbled by New Mexican silver and turquoise rings, dreamcatcher earrings snagging in

> *long hair, notebooks and pencils and copies of* Black Elk Speaks *waving in outstretched hands, running after the truck?*

According to traditional stories of Old Chenega, humans turn into killer whales when they die. When killer whales come close to a village, the people believe they're calling someone. When friends scattered the ashes of Mike Bigg, the first non-Native killer whale expert in British Columbia, into the water off Vancouver Island, a pod of killer whales swam over.

> *If I'm going to reach you, is the best idea to drop this letter into the water, into the midst of a killer whale group the next time I'm in Prince William Sound? We named one of the seal-hunting transients after you. We didn't see him last year and thought he'd died, because a dead male transient washed up on an island last spring. But then, to our relief, he showed up again this summer, swimming with his old group, the survivors, the ones I was studying when I met you.*
>
> *The Chenega elders are dying, one by one: Eddie, you, your best friend Chenega Pete. Few people in the village hunt seals anymore. The transients are dying, one by one. Neither group can be replaced. If landscape helps define who we are, do we help define the landscape? What is the Sound without you in your skiff, hunting seals in the old spots? What is the Sound without the transients, hunting seals in the old spots? I'd like to think you're swimming out there, though it's very unscientific of me.*

I push the buttons of other killer whale researcher friends, not because I'm interested in Native knowledge and mythology of killer whales—we're all interested—but because they think I privilege it, hold it above the motivations and methods of Western science, of non-Native people living on the coast. *Natives are no better than anyone else,* they say. *They waged raids on other villages. Some, like the Tlingit, took slaves, lived under elaborate caste systems. They caused local extinctions of animals. They want traditional hunting rights, yet now they use snow machines and outboard motors.*

It's true that in basic ways, we're all the same. Animals ourselves, we're members of the same species, evolved over two million years from a common ancestor. We're built out of the same nucleotide pairs. Our elemental lineage can ultimately be traced back to oxygen, carbon, nitrogen, and hydrogen formed during the creation of the cosmos, released from explosions of collapsing stars. And the old stories show that we're prone to the same weaknesses: pride, greed, jealousy, revenge.

Nonetheless, we're more than the elements that make us up, more than genes. Our minds are formed by education, language, family, society, experience, tradition, and particularly, I think, landscape. Alaska Native people have been living on this coast for thousands of years, since, as they say, "time immemorial." Compared to the rest of us, their ties to this place are old. Killer whales have been swimming in the ocean in relatively the same forms for the last five million years. Compared to humans, whales and their ties to the sea are ancient.

Dear Mike,

If only I could have taken you on the boat with me, followed transients with you, heard your comments when the whales were hunting seals or killing them. I want that knowledge, that connection, but I have to create my own belief system from what I see, what I hear, what I read. Is it always this individual, the work of the imagination? In your culture, was collective imagination passed down by the storytellers? As a scientist, I tell stories translated into science's language, so they will not be dismissed as "anecdotes." As a writer, I strive to write stories to express my evolving worldview, even if it's nearsighted, hindered, incomplete.

You told me how drastic the seal decline in the Sound was, and you blamed us, all of us, Native and non-Native. The planet is changing. We keep collecting data. We keep writing papers. But it's not enough. We've got to start telling the important stories, to give them meaning. It's becoming a matter of survival.

I'll practice by telling a story. Craig narrated the bones of it to me one day last week. It's the scientific creation story of the killer whale. Let me tell it as elder Willie Marks told the Tlingit killer whale origin story in the anthology *Haa Shuka, Our Ancestors*. It's poetry, his telling, with its pauses and repetitions. And it's a transformation story. Paleontologists believe that the evolution of whales represents the most complete transformation undergone by any mammal.

Once, in the distant time,
53 million years ago,
the ancestors of killer whales lived on land.
They lived on land,
those hippo-ancestors of killer whale.
Over millions of years, they gradually lost
those hippo-like limbs.
This happened in the ancient Tethys Sea,

where Pakistan is now. Some of this land
is now at the top of the Himalayas.
That's where you'll find
those sediments today.
Anthracotheres, they were called,
the proto–killer whales, medium sized,
piggish beasts
with four-hooved toes
on each front foot.
They had hooves, to walk about.
Then came Pakicetus. They transformed
into Pakicetus, furry, small, meat-eaters
with hooves. They moved into
the river channels and swam around like otters.
Then came Ambulocetus, they transformed
into Ambulocetus, the walking, swimming whales,
with thick, splayed out legs, four toes with little
hooves on the end, they were the size of
large sea lions, those proto-whales,
but they looked and acted like furry crocodiles
who lay half-submerged
in shallow water, waiting to ambush
their prey. Predators, they were,
stealthy hunters.
Then came Rodocetus, they transformed
into Rodocetus, with reduced hind limbs.
Rodocetus had flukes! No more
could they go walking
easily around on land.
And then came Zueglodon, they transformed
into Zueglodon, fish and shark eaters
with even smaller hind limbs.
They'd lost their body hair now.
Now they were long and snake-like
with small heads.
How could killer whales come from those?
But then came Squallodonts, they transformed
into the Squallodonts,
which look like the beaked whales of today,

but dolphin-sized.
They lived in the ancient time,
6 to 25 million years ago. They were carnivores.
Meat-eaters with robust teeth.
They probably lived like the killer whales
but they died out.
Something wasn't right.
The Squallodonts died out.
Now the fossil record,
the ancient origin story of killer whales,
it gets sketchy around that distant time
when the whales separated.
Some went the toothed whale way.
Some went the baleen whale way.
Some began to grow baleen.
And you can see this,
that they were once
connected, because even today, baleen whales
grow teeth in the womb, and the fetal
body reabsorbs them again. As a reminder.
They were once connected.
The way we have gills
when we're in the womb.
To show that we are connected.
And 12 million years ago came the great radiation.
And many species evolved.
Because the ocean in Miocene times
had many prey. And different ways of life
emerged. The Kentriodontids evolved.
The early killer whale ancestors transformed
into the Kentriodontids. The primitive dolphins.
And five million years ago, all the whales
as we know them now were here.
They were all here. Pretty much
the same as they are now.
Five million years ago.
And our earliest ancestors,
those proto-humans,
they didn't show up

until two million years ago.
The killer whales, they are much
older than we are. The killer whales
are ancient. They have been
living here a long time.
Maybe there is something
they can teach us
about living here.

Through their transformations, killer whales took the forms of the animals they would eventually prey upon. Maybe they carry a glimmer of this in their cells, the way they still carry finger bones in their flippers, reminders of their origins, so they can put themselves in the place of their prey, knowing their ways. And we carry this knowledge, too.

The ancient stories—legends we call them—taught right speech, right action, consequences. How to live as a member of society. To be humble, but not stupid. To hunt correctly. To know what actions turn animals away. To stay true to your place in the local ecology. Can science teach us this? Aren't the stories science needs to tell the ones that teach us where we went wrong, how to find our place again?

A well-known Tlingit and Tsimshian story recalls the origin of the *Keet Shagoon*, the killer whale. Willie Marks told Nora Dauenhauer one version of the story in 1972 in Juneau, when he was seventy years old, and she transcribed it in *Haa Shuka, Our Ancestors*. It goes something like this:

There was a great hunter named Naatsilane who was tricked onto a sea lion rock and then abandoned by his brothers-in-law because they were jealous of his hunting abilities. Naatsilane was a show-off. After being stuck on that rock for hours, a loon showed him the way to enter the sea, lifting it up like a cloth. Naatsilane swam down and found the house of the sea lion people. Because he was human, he could see a spearhead lodged in the sick sea lion chief's body and removed it, and in turn, the sea lion people placed him in a big balloon, the container for the southwest wind, told him how to get home. On his journey, he carried in his mind two things. One was an image he'd seen above the sea lion chief's bed, a picture of a tall-finned, ferocious creature. Second, he carried the sea lion peoples' instruction, *to get to shore, think only of where you are going, not of this place*. When he faltered in his thoughts, he came back to the sea lion rock. He tried again and made it to a beach near his village. With food and tools secreted to him by his wife, who'd been told by her brothers that he'd died in an accident, he hid and carved ferocious creatures out of wood, remembering the images he'd seen in the sea lion lodge. His first attempts failed. When he launched the wooden killer whales, they flailed around, floated back to shore, no spirit in them. After many attempts, using different kinds of wood, he finally tried yellow cedar, and the whales lived and brought back seals and halibut in their mouths. He instructed the killer whales to attack his brothers-in-law when they went out in their boats, but to spare the youngest, who'd tried to save him. The killer whales did as they were told, killed the brothers-in-law, and Naatsilane ordered them never to kill humans again. But Naatsilane, he went into the forest, as Willie Marks told it, "maybe to wherever he would die."

Dear Mike,

The writer Gary Holthaus constructed this totem pole:

wisdom
knowledge
information
facts
data

Of it, he said, "Data is clearly low on this totem pole, but it has become a huge totem of its own in our culture." I want to understand this, to try to put what it means into sentences, but talking to you is like writing a poem. Too many words turn into stew. Writing to you helps me stay humble because I know you'd laugh if I got too "out there."

I miss that about you.

E.

Whenever I made a recording of killer whale calls, I collected data in a notebook: time, behavior category, counter number on the tape recorder, time the observation period ended. To be more systematic, each observation period would have been the same length. But I didn't want to miss anything. If five minutes were over and those mostly silent whales were calling, I didn't want to have to press stop on the recorder. If they attacked a seal in the middle of a recording session, I wanted to race up to see what was happening. I was impatient, exuberant, overly excited, so sometimes I was unsystematic. More naturalist than scientist, I wanted to know everything.

Back at the university, I entered my data onto spreadsheets, created number columns, tallied totals and means and standard deviations. What are stacks of numbers? What do data mean? To state facts, the scientist must interpret the data, look for verifiable patterns. From my data, facts emerged. The AT1 transients were mostly silent when foraging for marine mammals and mostly noisy after killing them. Many facts taken together become information about a species, and the accumulation of information, some contradictory, increases our knowledge. But the step to wisdom is less certain.

The story of Naatsilane is derived from data, facts, information, knowledge. The first people on this coast, and people like Mike, accumulated data and established facts while hunting, watching, traveling, gathering, exploring, surviving. Naatsilane's story endured because it contained wisdom, and wisdom has survival value. It remains relevant over time. It's something very different from knowledge. Scientists categorize stories such as Naatsilane's as myths, and yet the factual nature of this story is emphasized by its tellers. This is not fiction. J. B. Fawcett, another elder whose version of *Keet Shagoon* Nora Dauenhauer recorded, said, "This one is true, this story. This is not a story without value." Wisdom, he implies, is knowledge that has value.

From the facts, people learn the origin story of the killer whale, and they derive knowledge of right action. The brothers-in-law were jealous, and they acted on it and were punished. But the wisdom in the story is more subtle. Although Naatsilane was the story's hero, who saved himself, who created killer whales, who successfully took revenge, his fate was isolation and death in the forest. And his bragging about his hunting prowess got him into trouble in the first place. In the end, he played the god. And this is a very old story with us.

And like any good story, different people pull different lessons from it. A scientist might notice that those sea lion people possessed a prescient image of the killer whale. They imagined an animal that would be their predator, and they imparted that image to Naatsilane, and so the story of *Keet Shagoon* illustrates the evolutionary connection between predator and prey—a contract, a dance—of signaling between sea lions and killer whales.

They appear to the south, from the direction of Matushka Island. We've been waiting here on the boat for hours, idling in the swells, watching sea lions, most of them sleeping, others sliding down the smooth rock slope with retreating swells, catapulting back up on rising crests, the sucking back leaving them high and dry. Pups and juveniles tousle in the churning shallows. The killer whales' respiration pattern is regular as they approach, with several synchronous breaths at the surface, bodies parallel, then several minutes below.

Next time they appear, they're within fifty meters of the rookery, and the mood on Chiswell Island changes. The resting sea lions heave themselves awake, arch their necks down toward the water, a collective roar moving through the colony every time a killer whale blows. Some swimming sea lions scramble up onto the rocks when the whales come close, but others group together, crane their heads and necks high out of the water, roar, then surge as one many-headed body toward the whales. After a few surfacings along the rookery and inside a cleavage on its northwestern side, the whales move away again to the north, rest, and then, after ten minutes, return for another variation of the dance. No kills this afternoon.

How many times has Matushka, the matriarch of the group, participated in this ritual with the sea lions? And who taught her the technique, the way she teaches her grandson, the calf, by bringing him close to the rookery, by helping him kill puffins and auklets for practice in the off time?

~

During my month-long writing residency in Sitka, in southeastern Alaska, Jan Straley, a humpback whale biologist, took me out on her boat to look for whales. One afternoon, as we photographed humpback flukes between snow squalls, she described her ambivalent feelings toward transients, who, in contrast to humpbacks, at times appear to torture their prey. She once watched a group of transients pin a humpback whale calf against shore, torment, then leave it. Another day, near Point Adolphus, six killer whales injured a sea lion. Its intestines streamed out of its body as it swam. A humpback whale whom she described as very odd, a kind of loner or outcast, followed the killer whales and injured sea lion as they swam most of the way to Hoonah, a Tlingit village on Chichagof Island. She's seen transients kill seabirds, mortally injuring them, leaving so rapidly she can hardly keep up in her skiff. Once, she and her husband encountered transients feeding. A large male swam under the boat with something red in its teeth. Jan said they should biopsy the whale, but her husband said, "No way. I would never be able to get rid of the bad luck. I won't touch those whales." Jan joked that he'd been listening to too many Tlingit legends.

In Inupiaq masks, I've sensed the blank realm of non-being, a nebulous boundary between the dead and the living. On Tlingit totem poles in Sitka, on the faces of creatures, I've sensed the thin boundary between the animal and the human, something cunning at once familiar and strange.

There's a missing, untold piece in science's published stories, the imaginative journey we're afraid to take, the questions we don't ask, which we allude to in private conversations amongst ourselves. Why did the injured sea lion swim with its predators to Hoonah? Why didn't the killer whales eat it? Why did the loner humpback whale join them? Can science answer? We shrug our shoulders, say, "Weird." When I study killer whale faces on totem poles, I see that the carvers recognized this weird aspect of killer whale—and human—nature, a quality that science can't penetrate or even describe.

According to the Tlingits, when the killer whale tribe starts north, the seals say, "Here comes another battle. Here come the warriors." And Jan said, laughing, "Why can't they stop being so mean?" Barry Lopez writes, "To allow mystery, which is to say to yourself, 'There could be more, there could be things we don't understand,' is not to damn knowledge . . . It is to permit yourself an extraordinary freedom: someone else does not have to be wrong in order that you may be right." He proposes that "if we are going to learn more about animals—real knowledge, not more facts— . . . we are going to have to find ways in which single, startling incidents in animal behavior, now discarded in the winnowing process of science's data assembly, can be preserved, can somehow be incorporated." The tool that can do this work of incorporation—transforming data into wisdom—is imagination. Its vehicle is story.

Dear Mike,

This evening in my flower garden, I dispersed wild iris seeds I collected from your grave site. The wild among the cultivated, the native among the tamed. The way our lives are nests within nests, our imagining selves in the empirical world, in the verifiable world of disease and treatment, of hammer and plank. The world of Prince William Sound, you in your skiff, a dead seal in the bow, stopping by Whale Camp to say here is my gun, here is a seal I killed, this is who I am. Predator like those predators I study, those other hunters you encountered in Icy Bay, Iktua Bay, around Danger Island. I don't know if irises sprout from seeds, but I'll take comfort from sowing these brown pods among the lily stalks.

I stand on the bow of *Whale 2* off New Year Island. We've cut the engine, dropped the hydrophone because these three transients are acting funny, high-arching, milling around in one place. We've finally learned to recognize this, after years of keeping our distance, scribbling in the field book, "Milling, silent." Now we know to get close fast, throw the hydrophone over, look in the mouths of the whales, sometimes even under their pectoral flippers, where they stash chunks of meat like schoolchildren packing textbooks. Scan the water for an oil slick spreading, hairs and blubber bits ascending from the green-black. Turn the recorder on to catch the transients' subtle squawks and bleats so easily masked by wave slaps. Listen close. Listen. And then she floats over, a big seal crosswise in her jaws, clouds of blood pulsing from the seal's body like lion's mane jellyfish, those big pulses the still-beating heart pumps out. And she swims beneath the boat where I am standing as if to say, *This is who I am.*

Dear Mike,

Today I got an e-mail from my friend Paul, who works "across the bay," as we say here in Homer, on the tamed side of Kachemak Bay. He told me that last year, in one of the villages, an elder lay dying, surrounded by his loved ones. He wouldn't pass. Someone spotted killer whales, a small group, deep in the bay. Transients, I thought. This person ran in and told the family, and they whispered in the dying man's ear, "The arllut are here, they've come for you," and he died.

Stories. We use them to make sense of the world. Today I read another kind of story, in a scientific article by Springer et al. Conventional wisdom has it that ecosystem shifts occur from the bottom up, that water temperature changes impact the phyto- and zooplankton, and the ripples spread upward. These scientists wonder what would happen if we thought of it the other way around. What happened when so many great whales were removed from the oceans by industrial whaling? In their paper, speculations bob like wooden killer whale carvings. They ask, what if great whales were once the principal prey of killer whales, and after they were hunted down, the killer whales turned to seals and sea lions for food? Will their theories wobble and float, or will they capsize? Like Naatsilane's carved creatures, will they have life? If I trace those theories to their source, I find stories. Some biologists in the Aleutian Islands saw killer whales enter a kelp bed where sea otters rested. A ruckus ensued and some otters disappeared. This is the place where . . . These are the whales who . . . And the story was passed on, turned around in the mind, woven with stories of whaling, of Antarctic killer whales, of gray whale kills, until this new story emerged. Are they shamans? What would you say? We want to understand, to know. It's stories, in the end. It's our human way. But you know that.

E.

Not all stories are true, in the end. Not all lead to wisdom. But we can learn from them. They can teach us something about ourselves, about where we go wrong.

When we look at killer whales or other predators, we recognize our own cunning, how the struggle to survive can be brutal. It's ironic that the predatory animals that still have the power to shake us, to scare us with their killing power and ruthless drive to live, are in mortal danger from us, killers of another order. In Silver Bay, in front of an abandoned pulp mill site near Sitka, where Jan has seen transients hunting, a cordoned-off mat of toxic waste floats near shore, too volatile to be moved. A story of where we went wrong.

Dear Mike,

I found this old journal entry this morning and thought of you. "I don't believe we go up when we die. We go under. Like Naatsilane, we slip beneath the cloth of the sea, the reflected clouds, to the forest below, a gull beating strokes through the leaves with firm upthrustings of wings, the forest with no outline, made of flickering ribbons of color. The inverted world. Where the true nature of the killer whale can be found."

Naatsilane went under the sea to find things out. Jan goes under the sea in a mechanical pod to watch humpbacks feeding. Other scientists go under in the Aleutian Islands, emerge with stories of forests of sea pens, sea peaches, gorgonian corals. On video footage, we hear how their voices, their empirical minds, give way like snow bridges over crevasses to awe. Writer Doug Chadwick says science is an "organized form of wonder." In Iceland, scientists went under in winter and saw killer whales trapping and concentrating herring schools with their shark-like circling, the loud cracks of tail slaps. They emerged with a metaphor: carousel feeding. The story is placed in a cloth-covered box, held close, brought to the table to be transformed into data, the knitting needles working wildly among the graduate students and professors: ecology, optimum foraging, predator/prey modeling. The natural laws that govern the world of hunger.

Our greatest hunger is for knowledge. But knowledge without a story to tell, without advocacy for the living world, is cold and still, like knowledge trapped inside a glacier. People wait for it, for the glacier to force the story to its face, where it becomes recognizable, where it can make us wiser.

In *Consilience*, E. O. Wilson describes his struggle to develop a synthesis of all knowledge—one unifying, scientifically based theory of everything that will explain the world and ourselves in it. I hated this notion when I first heard of it. Science is a giant net. Strands are woven and knotted together out of logic, data, deductive reasoning. Some animal rudely busts through the net, some experiment goes awry, and the whole fabric is compromised. The net must be re-visioned, reformed. This wiley critter must be examined from behind a barrier that keeps us removed. Like blind people, we feel every strand, every joint, kept always from the knowledge we seek by the net itself.

And science has created a net of words, a language that also keeps us in our place, the animal in its place. Once I believed that if I studied hard enough, I could finally see that net, grasp it, know its laws, learn its language so I could ask the rightly phrased questions. Being out in the field day after day, year after year, I've come to see it's not enough to understand the world through the net of science, not for me. But my scientifically trained mind so colors the way I perceive nature that I can't even see what I'm up against.

This morning, I found that book, *Consilience: The Unity of Knowledge*, on our bookshelf. Rereading, it struck me that what I at first saw as arrogant was in fact creative. It represented the mind of science meeting the mind of imagination in an attempt to integrate our knowledge with our living. We try to connect threads to make a story, to build a net with meaning. The scientist uses a particular loom; the Sugpiaq hunter uses another.

As researchers, we write up our findings, and in the discussion section, try to fit our results into current ecological or behavioral theory. Not just this animal's story, this pod's story, this population's story, this species' story, but the story of predation, the story of group living, of population dynamics: unifying stories. It's something that connects us as a species. We're net-makers. Here's a color of thread no one knows the name of. It didn't drop out of the sky, out of a black hole, from a passing UFO. It's a fact of the physical world. We tie it in. Here's a story of the origin of the killer whale. Here's a story of the demise of sea lions linked to decimation of great whale populations by commercial slaughter. Here's the rebuttal of the story by other tellers. They look at the web, say no, you missed this strand, that knot. We mend the net. Here's a story of how killer whales ended up in Alaska in four different forms. Here's the story of what happens when we die.

Native people, in traditional times, felt, like E. O. Wilson, that the world was governed by natural laws. Living close to animals, to nature, observing the world at work, they in turn divined spiritual law: how to behave, how to act, how to live. Much as we do. What is the value of species diversity? How do we act toward our planet, toward each other? How do we reduce suffering? How do we understand who we are and what we mean? Is this the work of science?

E. O. Wilson admits that "people need a sacred narrative. They must have a sense of larger purpose, in one form or another, however intellectualized . . . They will find a way to keep the ancestral spirits alive." Can science be a part of that?

Dear Mike,

I dreamed I was at Whale Camp at dusk and heard puffs sounding like sea lions. I saw fins along the shore. I shouted to everyone, but it was much too dark to photograph. The whales came to the beach, and I ran to the edge. They had strange, pointy, nicked fins, a kind of killer whale we'd never seen before. My stepson Lars yelled that they were the kind that took mammals from the beach. "Get away!" he shouted. I tried but couldn't move my legs, and those gray-skinned whales swam parallel to me, and there was a battle inside me, whether I should be afraid or not.

There's something I'm scared of—that all the hours I spent studying transient killer whales, learning their ways, analyzing data, performing statistical tests, summarizing, presenting, won't make a damn bit of difference. We're going to lose them. I'm scared of that silence after they're gone, of the empty coves and passages. What would you tell me? Do we have to lose many more things before enough stories accumulate to turn us around?

There's a split in Sugpiaq stories about killer whales. Many Chenega people who responded to a survey I circulated said, emphatically, that the whales were good luck. They drove salmon into the area. A woman from Nanwalek said that sighting killer whales in July means the red salmon are coming, and people celebrate. She said also, though, that people have been "traditionally leery of them. We are supposed to share snuff with them." Herman Moonin, of English Bay, wrote that people respected the whales, didn't bother them, but sometimes asked whales to help with hunting, and the whales would scare seals or sea lions in their direction. He described how the people would talk to the killer whales through an oar, and from far away the killer whales would come and float next to the hunters' boats.

Moonin told a story about a man, who, in modern times, might very well have been called a scientist:

> There was a man one time that wanted to see if it was true that killer whales bring game to hunters. So he went hunting in his baidarka skin boat and took some snuff to the whales, "Killer whales, come and have some snuff." They came to him with their mouths open, and they had their lips and teeth apart where he was going to put the snuff in their mouths . . . The whales would take the snuff and dive under the water, then other killer whales would come up for snuff. When he was out of snuff, the killer whales left and swam into a cove.
>
> This man was very curious about the whales, so he followed them into the cove . . . It was muddy at the head of the cove and when he got there, he saw tracks. The tracks that he saw were like humans' but different. Some were real small and some were huge. They all headed inland toward the same direction. The hunter thought to himself, "Killer whales must really be people that put on killer whale clothes."

Moonin then recounts that very long ago, when someone died, the killer whales would come to take them to a certain cove, dress them like killer whales, and release them into their new form. According to this story, the only difference between killer whales and humans is our skins. Zipping and unzipping this skin is like lifting up the cloth of the sea to go under, to effortlessly enter the killer whale realm. It seems magical, this lifting of cloth, this zipping on of skin. But it's much like the evolution story, in which killer whales shed body shapes to become what they've been now for five million years. Killer whales know some things about living here. Maybe we have to shed the skins we've been wearing, find our way back into the weave, rejoin the ecosystem, put back on our animal skins, but it isn't simple.

Last spring, I attended a yearly ritual on the Homer Spit, a reenactment of Sugpiaq hunters returning from forays. As the skin boats approached the beach, a big seal circled the bight and approached the kayaks, lifting its head, not recognizing them as predators. Like other skins, fog strips stretched loosely over the bay, mercuric milk in which some fishermen waded, casting. Pink surveyor's tape marked the landing spot. A dozer slumbered at the high-tide line. Anchored to shore, a steel fishing boat waited for the flood. Children in fox tails and skins beat drums, moved their arms in welcome. The weather was down, so planes from Nanwalek, across the Inlet, carrying dancers and the Russian priest, were grounded.

So the paddlers beached, and no one knew what to do. Finally, a museum curator with a megaphone walked an elder to the water's edge to welcome and bless the paddlers. After the Our Father, we stepped down the beach to steady the boats as barefoot hunters disembarked, stood smiling. We hovered, uncertain and complicit. I kept thinking of the way I'd trained my binoculars carefully in the direction of Iktua Rocks the previous summer because years ago, I often found hunting transients there. I'd tried to return not only to a place, but to a time, recognizable by stories that flickered against the water and islands, whales and no whales, seals and no seals, hunters and no hunters.

A woman from Dolovan, near Nome, told me of a time that killer whales helped her people find food. When she was a baby, her family was moved from Elim to Dolovan. Some people went overland. Her grandmother and others went by rowboat around Cape Darby, in the Bering Sea. She herself was in the boat, wrapped up in a rabbit-skin parka. The people were hungry and cold, so someone called to the killer whales and asked them for food. The next day, big pieces of muktuk washed up on the beach. The people ate it raw, they were so hungry, and the oil stained their clothes, which had to be burned.

"You never play with or harm or hunt or harass a killer whale," she said, "because they are so close to people." She told me that a woman in Dolovan married a white man who didn't know all of the traditional rituals or rules, and one day he shot a baby killer whale. "A person who harms a killer whale will die," she said. An adult killer whale showed up and started swimming through the bay back and forth. The white man finally confessed to his wife what he'd done. She blamed herself for failing to teach him properly, so she went to a point far out in the water and apologized to the killer whale, saying that her husband didn't know, that it was her fault. The whale eventually forgave them and left.

Inupiaq people say that killer whales drove seals onto the ice for hunters to catch. Tobacco was thrown into the whales' open mouths, in thanks. Those stories from many places in coastal Alaska, of killer whales open mouthed, lips pulled back, revealing their teeth to hunters in boats, remind me of Matushka. We first saw her in Prince William Sound in 1987 with some of her relatives on my first day volunteering on the research project with Craig.

While some of the whales swam rapidly under and around us, Matushka breached and tail-slapped repeatedly within a few meters of the skiff, dousing us with water. I was twenty-three and naive, didn't know that this wasn't ordinary killer whale behavior, so I screamed and jumped around and tried to touch her. Finally, I looked at Craig, salt water dripping from his beard, and saw his unease. It was weird, he said, for transients to interact with a boat this way. We couldn't even take identification photos for fear of ruining the camera, but more so, because the whales were too close. We finally had to back away from them, but they charged after.

That was my initiation into killer whale research, and I see it now as both a welcome and a warning, a warning that my stories would have to change. My imagination would have to expand to include Matushka as she glided along the hull of the boat, her mouth wide open, showing me her teeth. I would have to look into my own animal nature.

It's not impossible to imagine killer whales and humans having once spoken the same language, interchanged body forms. We are still dependent on each other, and the stories tell us we must act that way, unless we want killer whales to exist only as mythical creatures, like the thunderbird, who, in one story, did battle with a killer whale, driving it into the sea, where it's lived to this day. Our big, imaginative brains define us. Deprived of the creatures who inspire our stories, will we be human? Or will we be proto-something else?

Dear Mike,

I'm reading an article called "Indigenous Science" by Jurgen Kremer. He says that "indigenous consciousness is potentially accessible to everybody who is willing to put in the work and live in this way. You can be native and out of your indigenous mind, and you can be European and in your native mind." Do you think that's true?

Just as language shapes our thoughts, the way we tell stories shapes the way we see, and the way we see—what we look at, the amount of time we spend on the water, in the woods—shapes our imaginations. Kremer asks, "What if we have established a big thought system at the foundation of which is one giant rationalization? What if we lack sufficient context to make much of our knowledge meaningful? What if we need to turn things upside-down?" Is that the difference between knowledge and wisdom? Is wisdom knowledge turned upside-down?

I write poetry these days, a craft that encourages the holding of opposing truths in the mind at the same time. While my logical mind grapples to reconcile the Tlingit story of the origin of the killer whale with the paleontological story, in my other mind, they coexist, both true, both essential.

Dear Mike,

Your words one day on the ferry dock, when you shushed my jabber, said, "Shhhh, listen, I hear something" were a challenge to me. I listened. I couldn't hear what you heard. But I knew that there was something out there. Like a poem, what you said was unparaphrasable but true, factual.

The killer whale totem poles in Sitka challenged me, too, saying, "*Shhhh, look.*" In them, I perceived the killer whale's true nature, the Matushka nature. Humans made them, so they reflect our nature too. They are the products of imagination, knowledge, spirit. They are wise.

The pole rises among tree trunks. The blowhole of the killer whale is a literal hole, a round, human mouth. A white circle on the killer whale's head is the mouth's human face, wide-eyed. The whale's mouth is partially open in a grimace or smile, like Matushka's, baring rows of teeth. The eyes are round and black and ringed with blue, the tongue extending to touch the head of the owl below, passing on knowledge.

If I, as a scientist, could allow myself to be inverted, to be swallowed by my study animal, like the curious raven in Chief Makari's story who was swallowed by a killer whale so he could know what was inside it, how would I be different? What knowledge could I emerge with, blinking, back into the daylight?

When I left the totem pole park that day, a raven cawed from the treetops above me, like one of the ancestors, noticing me. Those spiritual certainties come in a flash, unbidden, then are drowned out by the logic of waking life. That's my divided nature, the divided nature of the Western scientist, cursed to expose spiritual feeling under the harsh light of intellectual scrutiny, my mind so thoroughly trained to skepticism that it's not even conscious. I want to move into that long-unexplored region, where many thought systems coalesce, as meaning does in a poem, where contradictions exist side by side, where things are never completely resolved.

All the while, cunning animal eyes glint with humor at my predicament, the raven perpetually following me, laughing from treetops or streetlamps. Matushka shadowing me, open-mouthed. You shadowing me in your skiff, chuckling. I want to tear through the net and, just once, see the world as it really is, as the raven sees, as Matushka sees, as you did and do.

Mike, I will try.

Acknowledgments

Several essays in this collection have been published, in different versions, in magazines and anthologies. The author gratefully acknowledges the following:

"Crossing the Entrance" (*Northwest Review*, January 2002).

"And Suddenly Nothing Happened" (*Connotations*, Spring 2001).

"Walking on Carlson Lake with Bill" (*Under Northern Lights*, ed. by Frank Soos and Kesler Woodward, University of Washington Press, 2000).

"Leaving Resurrection Bay" (*Prairie Schooner*, Fall 1999; reprinted in *American Nature Writing 2000*, ed. by John Murray, Oregon State University Press).

"Ghosts of the Island" (*Prairie Schooner*, Fall 1998).

"The Burden of the Beach" (*American Nature Writing 1996*, ed. by John Murray, Sierra Club Books, reprinted in *Intimate Nature*, ed. by Linda Hogan, Brenda Petersen and Deena Metzger, Fawcett-Columbine Press).

I thank the editors named above and additionally Hilda Raz, Ladette Randolph, John Witte, Carolyn Servid, and Dorik Mechau. The essay "Letters to Mike" was originally written for and read at the Sitka Whale Fest in Sitka, Alaska; thank you to Jan Straley for inviting it into the world. This book was a finalist for the Tupelo Press non-fiction prize in 2002, and I thank editor Jeffrey Levine for his encouragement. I am grateful to Marisa Farretto for generously allowing me to use a detail of her painting as this book's cover illustration, and to Michael Rosas-Walsh for photographing the painting. Deepest thanks also to Karl Becker for illustrating several of these essays.

In some of these essays, the names of people and places were changed to protect their privacy. Thanks to the landscapes and animals who populate this book, especially Prince William Sound and its killer whales, and to Molly Lou Freeman; John Lyle; Jon Miller; Lou Brown; Laurie Daniel; Kathy Turco; Olga von Ziegesar; Celia Hunter; Ginny Wood; Kyra and Neil Wagner; Mara, Asja, John, and Emily Saulitis; Carol Klumpp; Bill Fuller; Jan Straley; Elli, Lars, Eve, and Craig Matkin; Lance Barrett-Lennard; Kathy Heise; Graeme Ellis; Suzanne and David Selin; Harold Kalve; David Grimes; the Stowell family; Fred Rungee; and Mike Eleshansky. Wildlife biologist Bruce Dale generously related the story of the Riley Creek wolves to me. Thanks to those who shared their traditional knowledge of killer whales, particularly Pete Kompkoff, Nancy Yeaton, Mike Eleshansky, and Sperry Ash. *Paldies* to my parents, Janis and Asja Saulitis, for filling my childhood with books and language, and to my siblings, especially to my sister Mara, who buoyed me through the ups and downs of making this book and living its stories. For playing and writing with me in the wilderness, I thank the children of my heart, Elli, Lars, and Eve Matkin. For reading this manuscript at various stages, I'm indebted to Frank Soos, Peggy Shumaker, John Morgan, Richard Carr, John Keeble, Sherry Simpson, Nancy Lord, Molly Lou Freeman, Margaret Baker, Nicole Stellon, Lisa Zatz, Susan McGinnis, Enrico Sassi, Kyra Wagner, and my fellow graduate students at the University of Alaska Fairbanks. The gift of time and space for writing and revision was provided by fellowships from the Island Institute, Sitka, Alaska; the Rasmuson Foundation; and a Connie Boochever Fellowship from the Alaska State Council on the Arts. For the trail I followed, I thank in particular John Haines, Seth Kantner, Frank Soos, Nancy Lord, Peggy Shumaker, Sherry Simpson, Richard Nelson, Carolyn Servid, and Marybeth Holleman. I am lucky to work with scientists whose curiosity and passion stem from love for and delight in the natural world and its inhabitants: Craig Matkin, Olga von Ziegesar, Bud Fay, Lance Barrett-Lennard, Kathy Heise, Graeme Ellis, Mike Bigg, John Ford, Jan Straley, Harald Yurk, Volker Deecke, and Dena Matkin. Thank you to Kate Gale and Mark Cull at Red Hen Press for their belief in this book and—with Peggy Shumaker—for their vision for Alaskan literature. In addition, I am indebted to Peggy, and to Joeth Zucco, for final editing of these essays. My partner Craig Matkin knew when I needed an anchor and when I needed freedom and space. A writer couldn't ask for more than that. I thank him for his patience and support. Finally, I'm blessed to have a friend like Sean McGuire, whose advocacy for the earth is my gold standard, and who said, "Eva, you are a writer," during a dark time. It made all the difference.

Notes and Credits

Poem excerpts by Molly Lou Freeman are used by permission of the author.

The epigraph for the book, by Anne Carson, is from "Kinds of Water," in *Plainwater: Essays and Poetry* (New York: Alfred A. Knopf, NY, 1995).

"The Burden of the Beach"

The Chugach stories referred to are from *Chugach Legends: Stories and Photographs of the Chugach Region*, edited by John F. C. Johnson and Barbara Page (Anchorage: Chugach Alaska Corporation, 1984).

Adrienne Rich is quoted from "Diving into the Wreck," in *Diving into the Wreck: Poems 1971–1972* (New York: W. W. Norton & Company, 1973).

"Somewhere Down That Crazy River"

Li-Young Lee is quoted from "Pillow" in *Book of My Nights: Poems* (Rochester, NY: BOA Editions, 2001).

Lyrics from "Somewhere Down the Crazy River" are by Robbie Robertson (Geffen Records, 1987).

"To the Reader"

Eli Wiesel is quoted from "Why I Write: Making No Become Yes," *The New York Times Book Review*, April 14, 1985.

"Looking for Gubbio"

The "dictionary of symbols" is J. C. Cooper's *An Illustrated Encyclopedia of Traditional Symbols* (London: Thames and Hudson, 1978).

Information about the wolf of Gubbio, the Sioux traveling song, and the wolf and killer whale legend were found in *Of Wolves and Men* by Barry Lopez (New York: Charles Scribner's Sons, 1978).

The story of the Riley Creek wolves was told to me by Bruce Dale, then a graduate student at the University of Alaska Fairbanks and currently a biologist with the Alaska Department of Fish and Game.

Rainer Maria Rilke quote is from *Duino Elegies* translated by David Young (New York: W. W. Norton & Company, 1992).

James Wright quote is from "The Jewel," from *The Branch Will Not Break* (Hanover, NH: Wesleyan University Press, 1963).

"Six Hundred and Fifty Pieces of Glass"

Rainer Maria Rilke quote is from "Tenth Elegy," in *Duino Elegies*, translated by David Young (New York: W. W. Norton & Company, 1992).

"Ghosts of the Island"

Excerpt by Mary Lavin is from "In the Middle of the Fields," in *Selected Stories* (New York: Penguin Books, 1984).

Details about the 1964 earthquake are from *The Day That Cries Forever: Stories of the Destruction of Chenega during the 1964 Alaska Earthquake*, edited by John E. Smelcer (Anchorage: Chenega Future Inc., 2006).

"Crossing the Entrance"

Excerpts from *Chapman's Piloting and Seamanship* are by Elbert S. Maloney, from the 61st edition (New York: Hearst Books, 2003).

Excerpts from *Coast Pilot 9* are from the 24th edition (Washington, D.C.: NOAA, 2006).

"One-Hundred-Hour Maintenance"

Adrienne Rich quote is from "Transcendental Etude," from *The Dream of a Common Language: 1974–1977* (New York: W. W. Norton & Company, 1993).

"Leaving Resurrection Bay"

Aleut sayings are from *Unangam Ungiikangin Kayux Tunusangin/Unangam Unikangis Ama Tunuzangis: Aleut Tales and Narratives*, edited by Knut Bergsland and Moses Dirks (University of Alaska Fairbanks: Alaska Native Language Center, 1990).

Sisiutl legend is from *Daughters of Copper Woman* by Anne Cameron (Vancouver, British Columbia: Press Gang Publishers, 1996).

Richard Nelson quote is from *The Island Within* (San Francisco: North Point Press, 1989).

Don Quixote quote is by Miguel de Cervantes, translated by Edith Grossman (New York: Ecco Press, 2005).

"Wondering Where the Whales Are"

Forrest Gander quotes are from "The Nymph Stick Insect: Observations on Science, Poetry, and Creation," from *A Faithful Existence: Reading, Memory, and Transcendence* (Emeryville, CA: Shoemaker & Hoard, 2005).

Adrienne Rich is quoted from "Diving into the Wreck."

"And Suddenly, Nothing Happened"

I must acknowledge my friend Christopher Schmidt for the title of this essay.

Rockwell Kent quote is from *Wilderness: A Journal of Quiet Adventure in Alaska* (Middleton, CT: Wesleyan University Press, 1996).

Gaston Bachelard quotes are from *The Poetics of Space*, translated by Maria Jolas (Boston: Beacon Press, 1994).

Jane Hirschfield quotes are from *Nine Gates: Entering the Mind of Poetry* (New York: Harper Collins, 1997).

Tuesdays with Morrie by Mitch Albom (New York: Random House, 2002).

"Seven Januaries"

Quoted haikus are from *The Essential Haiku*, edited and translated by Robert Hass (Hopewell, NJ: The Ecco Press, 1994).

Gary Snyder's "Earth Verse" is from *Mountains and Rivers Without End* (Washington, D.C.: Counterpoint Press, 1997).

"Epilogue: Letters To Mike"

Information about Tlingit oral narratives is from *Haa Shuka, Our Ancestors*, edited by Nora Marks and Richard Dauenhauer (Seattle: University of Washington Press, 1987).

Gary Holthaus quote is from "Ways of Seeing, Ways of Knowing," in *Connotations* (Summer 1997).

Jurgen Kremer quote is from "Indigenous Science: Introduction," in *ReVision* 18, no. 3, (Winter 1996).

Barry Lopez is quoted from *Of Wolves and Men* (New York: Charles Scribner's Sons, 1978).

Doug Chadwick quote is from *The Beast the Color of Winter: The Mountain Goat Observed* (Lincoln: University of Nebraska Press, 2002).

Quotes from E. O. Wilson are from *Consilience: The Unity of Knowledge* (New York: Vintage Books, 1999).

Henry Moonin story is from "Killer Whale Arllut," printed in the newsletter *Alexandrovsk: English Bay in its Traditional Ways* (English Bay School, Alaska).

Biographical Note

Trained as a marine biologist, Eva Saulitis has spent twenty-one years studying the killer whales of Prince William Sound, Alaska with her partner, Craig Matkin. In 1999, she received her MFA in creative writing from the University of Alaska Fairbanks, and since that time, her poems and essays have appeared in numerous magazines and anthologies. As a contributor to *Homeground: Language for an American Landscape*, edited by Barry Lopez, she has read her work on the PBS radio series *Living on Earth*. She spends several weeks each summer on Prince William Sound and Kenai Fjords, aboard the research vessel *Natoa*, and winters in Homer, Alaska, where she teaches English and creative writing at the Kachemak Bay branch of the University of Alaska.